The Earth's Last Wilderness

Broadway Books   New York

The Earth's Last Wilderness

A Quest to Save Antarctica

Robert Swan with Gil Reavill

Originally published as
Antarctica 2041

BROADWAY

BROADWAY BOOKS and the Broadway Books colophon are trademarks of
Random House, Inc.

Originally published in hardcover in the United States as *Antarctica 2041:
My Quest to Save the Earth's Last Wilderness* by Broadway Books,
an imprint of the Crown Publishing Group, a division of Random House, Inc.,
New York, in 2009.

Library of Congress Cataloging-in-Publication Data

Swan, Robert.
The earth's last wilderness: A quest to save Antarctica / Robert Swan,
with Gil Reavill.—1st ed.
p.   cm.
1. Antarctica—Environmental conditions.   2. Environmental protection—Antarctica.
3. Climatic changes—Environmental aspects—Antarctica.   4. Swan, Robert.
I. Reavill, Gil, 1953–   II. Title.   III. Title: Antarctica twenty forty-one.
GE160.A6S93 2009
577.5'8609989—dc22   2009010963

ISBN 978-0-7679-3176-2

Printed in the United States of America

Design by Leonard W. Henderson
Map graphics by Susan Lopeman
Photographs courtesy of the author

10 9 8 7 6 5 4 3 2 1

First Paperback Edition

This book is dedicated to the generations
that came before and the generations to come after—

Margaret "Em" Swan,
age ninety-four,
and Barney Swan,
age fifteen

And to Peter Austin Malcolm (1956–2009),
an Immortal among the Immortals—
without him our dream would never
have taken shape

We are all adventurers here, I suppose, and wild doings in wild countries appeal to us as nothing else could do. It is good to know that there remain wild corners of this dreadfully civilized world.
—ROBERT F. SCOTT

The world ought to have the sense to leave just one place on earth alone—Antarctica.
—PETER SCOTT

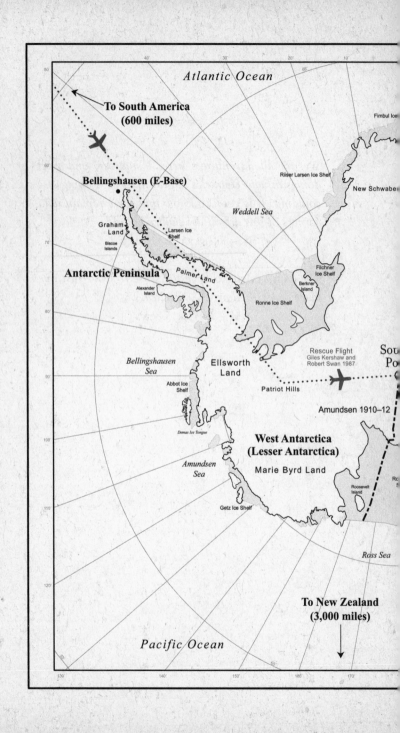

Atlantic Ocean

To South America
(600 miles)

Fimbul Ice

Rüser Larsen Ice Shelf

New Schwabe

Bellingshausen (E-Base)

Weddell Sea

Graham
Land

Larsen Ice
Shelf

Biscoe
Islands

Filchner
Ice Shelf

Antarctic Peninsula

Palmer Land

Berkner
Island

Alexander
Island

Ronne Ice Shelf

Bellingshausen
Sea

Ellsworth
Land

Rescue Flight
Giles Kershaw and
Robert Swan 1987

Sou
Po

Abbot Ice
Shelf

Patriot Hills

Amundsen 1910–12

Demas Ice Tongue

**West Antarctica
(Lesser Antarctica)**

Amundsen
Sea

Marie Byrd Land

Ro

Roosevelt
Island

Getz Ice Shelf

Ross Sea

**To New Zealand
(3,000 miles)**

Pacific Ocean

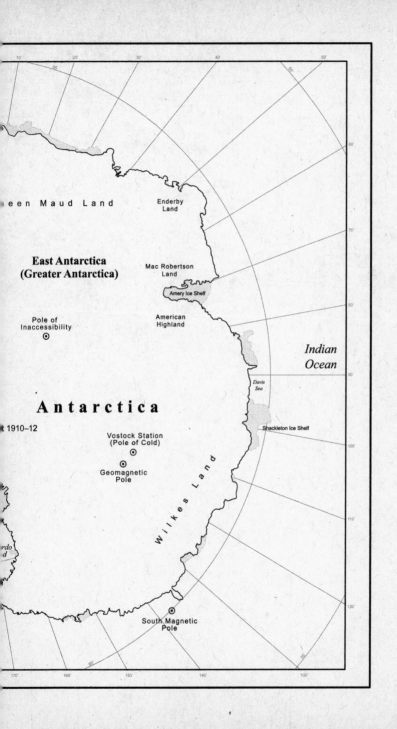

en Maud Land

Enderby
Land

**East Antarctica
(Greater Antarctica)**

Mac Robertson
Land

Amery Ice Shelf

American
Highland

Pole of
Inaccessibility
⊙

Indian
Ocean

Antarctica

Davis
Sea

Shackleton Ice Shelf

t 1910–12

Vostock Station
(Pole of Cold)
⊙

⊙
Geomagnetic
Pole

W i l k e s   L a n d

rdo
d

South Magnetic
Pole
⊙

The Beardmore

In the Antarctic summer of 1985 I found myself standing at the inland margin of the Ross Ice Shelf, a crevasse-riven, glacier-fed formation about the size of France. A France without baguettes and cathedrals. A totally Paris-less France.

The ice beneath me ran down a thousand feet. Underneath that, the Bible-black darkness of a cold, unexplored sea.

There were reasons why the Ross Sea remained unexplored. A New Zealand fishing boat once pulled from its waters a colossal squid (*Mesonychoteuthis hamiltoni,* to distinguish the species from its smaller cousin, the merely giant squid) more than thirty feet long and weighing over a thousand pounds. That's what they had down there, that and God knows what other creatures. Perhaps only Captain Nemo could have handled it.

To report my location on the Ross Ice Shelf above the Ross Sea, in other words, is another way of saying that I was in the middle of frozen nowhere, perched on the brink of an enormous nothingness. "A silence deep with a breath like sleep" is how one man who died there put it.

Early Antarctic explorers called the ice shelf "the Great Ice Barrier," in honor of the hundred-foot-high vertical wall where it meets the sea. But for those early explorers, and for me, the barrier acted more as a road, an immense, human-dwarfing, windswept road, but nevertheless a well-recognized path into the interior of the continent.

We followed "In the Footsteps of Scott," as our expedition was called, tracing the trek to the South Pole of the great British explorer Captain Robert F. Scott.

As I stood at the edge of the barrier, the question I pondered was pretty basic. Why in heaven's name would anyone *want* to enter the interior of Antarctica? Why would anyone freely choose to experience the most inhuman landscape available to us? It's a boogeyman of a place, good for scaring the wits out of homebodies.

"The highest, driest, windiest, coldest place on earth"—that's a formulation that seems by law to appear in every single piece of writing about Antarctica (the most memorable example: a book coauthored by Leonardo DiCaprio and Mikhail Gorbachev).

No one had even laid eyes on the place until 1820. No one wintered over for a full year until 1898.

Why was I there?

With my thoughts jangled and my inner clock going haywire since the midnight sun rendered night into day, I realized I had no answer to that basic question. No answer at all.

I was twenty-nine years old. With two other team members, Roger Mear and Gareth Wood, I had just spent four weeks walking—trudging, struggling, sledging all our supplies ourselves—across the Ross Ice Shelf. We were ants on an ice cube the size of France.

Our ultimate goal, that of reaching the South Pole, now appeared a vain hope. By our calculations we were fifty-eight days away from 90 degrees south. Our food supply—biscuits, sausage, soup dosed with vegetable oil for added calories—was good for another fifty-five days.

We were already starving. Since we hauled our own provisions for the journey, we had calculated our supply down to the gram: 5,200 calories per day per man. It was not enough. As we labored over the ice cube, our bodies were eating themselves. My weight plummeted: the "South Polar diet," we called it.

Controlled lab research has demonstrated not only physiological but also psychological effects of semi-starvation: a tendency on the part of the hungry toward the so-called neurotic triad—hypochondria, depression, and hysteria. The triad hit me hard. I imagined symptoms, felt listless and low, and experienced periodic spikes of panic over our situation. We gave off the characteristic ketone odor of the starving vertebrate. I could smell it on myself as I wandered a short distance from the camp, late on that bright-as-day evening on the barrier.

South of us, barring our way to the pole, towered the twelve-thousand-foot peaks of the Queen Alexandra mountains. The range braced the mighty Beardmore Glacier, over a hundred miles long, the second-largest glacier in the world. To get to the pole, we would have to climb the Beardmore's immense "ladders of smashed glass," as it was described by Ernest Shackleton, the man who discovered it.

In the gross emptiness of the ice shelf, our camp appeared puny and insignificant, a tent with three ice sledges dumped over next to it. Robert F. Scott called this place, where he was nailed by a ten-day blizzard, "the Slough of Despond." "Miserable, utterly miserable," he wrote in his journal. "Slough" refers to an allegorical place of despair in *The Pilgrim's Progress*. The name joins other colorful waypoints on the path to the pole, such as Shambles Camp, Devil's Ballroom, and Butcher's Shop.

Our expedition had broken down. Harsh conditions, personality clashes, and, in my case, devastating self-doubt left us marooned in the most inhospitable environment on the planet. I knew by now that Roger Mear and Gareth Wood were hardly speaking to each other. The only thing they seemed to agree on was that they didn't much like me.

A half decade of planning, raising funds, and untangling problems of supply and transport had come down to this—three squabbling humans just a couple of arguments away from freezing to death. There was no hope of rescue. We had no radios. We were unaware of any human presence within four hundred miles of us. At that moment I was absolutely, positively convinced that my life had ended. More distressing still—if that's possible—my dream had ended.

I had scratched and clawed my way to this point. I had buttonholed famous mentors and explorer-scions such as Sir Peter Scott, Lord Edward Shackleton, Sir Vivian Fuchs, Lord Hunt, the Royal Geographic Society, and Jacques Cousteau. I had borrowed enormous sums on the basis of no more than my smile (my smile was in debt to the tune of $1.2 million). Then I slogged on ice-numbed feet to this forlorn point in front of the Beardmore, all in an effort to honor Scott and Shackleton, my boyhood heroes, by walking to the South Pole.

Why? For what purpose? To fail? To die?

Roger Mear was one of the world's foremost mountaineers. Gareth Wood, also an accomplished mountaineer, was a meticulous organizer and logistics whiz. What was I? I was a novice. I wasn't a mountaineer or even an outdoorsman. And I was a foolhardy novice. I had mounted an expedition to the South Pole without ever having really been camping before.

Standing out there in the frozen nowhere, we somehow had to find within ourselves the skills to sort out our difficulties and make the dream real again. We had to trek the remaining five hundred miles to our goal—tantamount to walking from the east coast of America to the Mississippi River, but on fissured, hillocky, dangerous ice.

We followed in the footsteps of Scott and his polar party. That meant we were following in the footsteps of death, since the whole party died on their return from the pole. On our trek we had already passed the spot where Scott himself and two of his mates perished, and the place where Captain Lawrence "Titus" Oates walked out into a blizzard in an act of self-sacrifice. Up ahead, at the foot of the Beardmore, was the death site of Taff Evans, the strongest of the group of five, and paradoxically the first to go.

Antarctica itself had it in mind to murder us. Antarctica the inhuman, Antarctica the hostile, Antarctica that cares not a whit whether humans live or die but obviously prefers them dead.

If we had dared to inhale an uncovered breath in some of the insane temperatures we encountered during the previous winter, our teeth could have possibly cracked and exploded like so many tiny artillery shells. In the depths of the sunless months, at a temperature of −80 degrees F, toss a pan of boiling water into the air, and it freezes with an odd crinkling sound before it hits the ground.

Bitter, lethally cold in winter. The lowest temperature recorded in Antarctica (−128.6 degrees F at Vostok in 1983) was 39 degrees colder than the lowest recorded temperature on any other continent.

The place had its warmer and fuzzier side, too. It's the only continent that has never hosted a shooting war. Antarctica has no verified homicides, no prisons, beyond petty theft no crime at all—an absence that, as far as I was concerned, merely rendered it all the more strange and inhuman.

It was relatively warm during our summer trek—we would hit a high one day of almost 40 degrees F—but still utterly alien to human life. Journeying to the continent was the closest anyone could get to leaving earth without actually resorting to space travel. Great beauty alternated with sheer terror. False suns hung in the sky, and false moons, too.

I had stared at only ice and snow for so long that I hallucinated shapes in the landscape wherever I looked: a Sioux chief in full feather headdress, the profile of Queen Victoria. In Antarctica, I often got the sensation that I was gazing at the most beautiful person on earth, right at the moment when the mask was pulled off to reveal a frightening monster.

I knew what kind of monster. This was the face of God. Not a kindly, patriarchal graybeard, either, but Spinoza's god, the cold, abstract, impersonal force of nature—not He but It, not Who but That, an ur-god churning out magic tricks, turning midnight into day or lighting up the heavens with the multicolored streamers of the aurora.

So I tried to make a deal. I begged. Isn't that one of the stages of the human experience of death? You deny, you rage, you weep, you try to bargain.

"Just don't kill us," I whispered. The winds that poured off the Beardmore took my words and whipped them instantly away. It was as though I was pleading with the whole continent. "Just don't kill us," I prayed to the monster. "Just don't kill us, and I promise I will somehow do whatever I can do to protect you."

Why 2041?

W ell, I was lying. At the very least, not telling the whole truth. Or maybe I just didn't really know what I was saying. I'm embarrassed to admit it now, but that promise I tossed into the polar winds rang false.

I had no intention, at that moment, of doing whatever I could do to protect Antarctica. I had no idea what that might even mean. It just sounded good as an offering to the polar gods. I probably would not have said it, or thought it, or whatever I did standing out there on the edge of the Ross Ice Shelf, had I known what it would actually entail.

It took the greater part of my life, but Antarctica would hold me to the oath I swore, to make every effort to protect it as the world's last remaining pristine wilderness. I spent years trying to duck that promise before I finally embraced it.

I want to leave myself marooned on the barrier in 1985 for a while, in order to relate the story of how we got there.

I am the first person in history to have walked to both the North and South poles. No matter where I go, that is the descriptive sentence that precedes me. I remember Sir Edmund Hillary, conqueror of Everest, cautioning me before I attempted my second pole walk. "Are you sure you want to do this?" he asked. "Once you've done it, it doesn't matter who you are or what else you do, the pole walks are going to come up. Are you sure you want to be married to it?"

Then he added, a shade wearily, spoken by a man who knew, "Because it gets a little tiresome after a while."

In my heart of hearts, I knew that the sentence should be amended with two words. I was the first person in history *stupid enough* to have walked to both poles. From the coast of McMurdo Sound, Antarctica, nine hundred miles to the South Pole. From Ellesmere Island, Canada, five hundred fifty miles to the North Pole.

I am a slow learner. It took almost every step of both those journeys to realize what that promise whispered to the Beardmore winds might entail. The greatest threat to the polar regions is human-induced climate change. Present-day, fossil-fuel-burning, carbon-emitting industrial civilization is burning the candle at both ends, burning the planet at both poles.

So two tasks complemented each other. We have to stop the ice from melting, and we have to prevent the Arctic and the Antarctic from being exploited. My promise to be an advocate for the polar regions, and for Antarctica in particular, meant that I had to join others in the fight against global warming, people such as Al Gore and Robert F. Kennedy, Jr., organizations such as World Wildlife Fund, Greenpeace, and the U.N.'s Intergovernmental Panel on Climate Change.

The promise is embodied in the title of this book and in the name of the organization I founded: 2041.

The year 2041 is when the international treaty protecting Antarctica begins to come up for review. It's the year when the fate of the last great wilderness on earth will be decided. The year 2041 first became important to me in that context.

But as the disaster of climate change loomed into sharper focus, 2041 took on a larger meaning. I began to register other mentions of midcentury trends and deadlines, and 2041 became a sort of signpost for a time when a number of cataclysmic environmental developments might converge. The year 2041 is when

- greenhouse-gas emissions, if they follow current trends, will rise to 700 gigatons per annum (700 billion tons annually), a level projected to induce a five-degree rise in average global temperature over the next century. Global warming will have become a reality, triggering extreme weather patterns, rising sea levels, and resource shortages that will cause widespread disruptions to life as we know it.

- given current use patterns and rates of increase in energy demands, global oil production will drop below twenty million barrels a day—the accepted level necessary for sustaining industrialized civilization.

- soot from the coal plants and "black carbon" from cooking stoves in China and India, falling on the surface of glaciers in the Himalayas, will cause them to absorb more rather than reflect sunlight, shrinking them by 75 percent and disrupting the water supply for billions of people.

- sea levels will have risen .5 meters, given current trends in accelerated glacier and ice-cap melt on the margins of Greenland and Antarctica. A half-meter rise renders untenable one-tenth of human shoreline habitation. For example, half the roadways in Cairns, Australia, will be underwater. Extreme sea level—the measurement of high seas during hurricanes and storm surges—will displace 200 million people and impact a fifth of the world's population, over 1 billion people.

- the last Alaskan polar bear will have starved to death in the wild—again, extrapolating from current trends, in this case of bear-habitat destruction and population decline. All in all, in 2041 extinction rates on earth will have approached an unimaginable threshold, with 1 million land-based species gone forever.

- the last of the snows made famous by Ernest Hemingway will have melted off Mount Kilimanjaro in Kenya, and the last of the glaciers will have disappeared from Montana's Glacier National Park.

I came to see 2041 as a time when, if we don't change our ways now, today, our lives and, more important, the lives of our children will have been irrevocably harmed. I wake up every morning with a number in my head, a ticking clock that measures the time that is elapsing until January 1, 2041. Ask me how many years we have until that date, and I'll be able to tell you. I have put a counter up on our 2041 website, tracking the days. The fuse is lit, the hourglass is turned, time is running out. It's like one of those digital timers on the bombs in action movies. *Tick, tick, tick.*

I confess that the whole 2041 concept is partly an inspirational tool. It marks the date not of an inevitable cataclysm but of a simple review of operations for an international treaty. Will the world suddenly blow up on 1/1/2041? Probably not. Am I a wild-haired doomsayer carrying around a placard reading THE END IS NEAR? I try not to be.

But throughout my years of working in leadership groups and focusing people on taking action, I've learned that it is invaluable to understand a task in terms of urgency and deadlines. That's how to get people off their duffs—by making them aware of the ticking clock.

Notice that all the predictions for 2041 are predicated on "if current trends continue." Hitting a wall in oil consumption, species decimation, sea levels, global temperature—all these outcomes are not inevitable. They can be addressed, changed, or averted by concerted international action.

On the other hand, a few climate-change activists and sober-minded scientists aren't talking in terms of 2041. They're talking 2020. I might be an optimist. We might not have three decades. We might have only one. The question is, When do changes to the biosphere—the place on the earth's surface where life dwells—take on an unstoppable momentum?

Where the crisis is clearest, where the threat is most immediate, is in the land that I love, Antarctica. That's what has led me to take up the challenge of climate change. To preserve Antarctica, we have to change the world.

The world's first oil industry did not start in Ohio or Pennsylvania or Texas. It started in New England, and its product was not petroleum pumped from subterranean cavities but oil extracted from the boiled blubber of bowhead, right, and baleen whales, sea mammals that were harvested at the point of a harpoon. Over the course of a century, the whaling ships had to go farther and farther to obtain the oil, as the near seas were hunted out.

Today we have been forced to go farther and farther—and deeper and deeper—for our petroleum. We are looking to the floor of the ocean and the frozen Arctic. It would be an unspeakable tragedy if our thirst for oil sent us all the way to the pristine reaches of Antarctica.

Imagine that we are working in the whale-oil industry in Nantucket in the nineteenth century. It was a vast and sprawling enterprise, providing light and energy (and whalebone corset stays) to the world. But the industry took the whales to the edge of extinction. As workers in the enterprise, we would have witnessed its total collapse over the course of just a few decades. Our Silicon Valley ceased to be. Our livelihood vanished.

Standing dumbfounded in once-bustling Nantucket, among the shuttered warehouses and ghost hulks of the whaling port, what would be our reaction? We might ask ourselves the perennial question of the unprepared and self-deceived: *What's next?*

Well, thanks to John D. Rockefeller and Standard Oil, we found a new source of energy. The new Nantucket was Houston, Texas, then Dubai and Brunei and Scotland. But we find ourselves on the down curve of the slope, in the dwindling phase of the market.

What's next? What has our modern-day Nantucket, the petroleum industry, given us? Great things, there's no argument about that. Central heating and light and transport, a rising standard of living, better health and cell phones and iPods. Our amazing contemporary world.

But really, what's next? The sooner we pose that question to ourselves, the sooner we grapple with it honestly and in good faith, the less likely we will find ourselves standing in a shattered modern-day Nantucket.

We need a new method of supplying ourselves with energy, and that method needs to take into account green concerns. It needs to be sustainable. It needs to supply us with light and heat and transport in such a way that it does not cook the atmosphere.

The gates to the last-discovered continent have opened wide. More people journeyed to Antarctica in the last decade than in all the years since it was first sighted. It's been only a century since Roald Amundsen, Robert F. Scott, and Ernest Shackleton made their epic expeditions of exploration. Could it be the case that climate change will have a devastating impact on a place that was first explored only a hundred years ago? Can humankind really screw things up just that fast?

To be able to look forward, it helps to look back. The first part of this book is an account of our expeditions to the South and North poles—how we did it, the obstacles along the way, the changes to my thinking because of my experiences. A backward look at what brought me to look forward, with urgency and hope, to the year 2041.

An integral part of my biographical history, and thus a second integral part of this book, is what transpired long before my time, during the history of discovery in the polar regions, now recognized as the Heroic Age of Antarctic Exploration. First and foremost, the endeavors of my personal triumvirate, Scott, Shackleton, and Amundsen, but also of those intrepid souls who came before them.

The year 2041 happens to mark the bicentennial of British naval officer James Clark Ross's charting of the Antarctic coastline in 1841, an expedition that I have always considered as the true beginning of the Heroic Age of Antarctic Exploration. Ross—who gave his name to the ice shelf I stood upon in 1985—has held a place in my pantheon of heroes ever since I was a child. Before the age of steam, commanding HMS *Erebus* and HMS *Terror*, two tublike former mortar-launching boats under sail, he displayed an astonishing vault-into-the-unknown bravery easily the equal of Columbus.

Before Ross, the map notations of the Antarctic Ocean read like this: "Many Isles & firm fields of Ice; Islands of Ice innumerable; Vast Mountains of Ice." This was *mare concretum*, the frozen sea, crusher and sinker of ships, a hazard that had halted every sailing captain before him.

Ross plunged forward. He entered the pack ice on January 4, 1841, and five days later he emerged into the open sea that would, like the ice shelf he discovered, later bear his name.

"Few people of the present day are capable of rightly appreciating this heroic deed, this brilliant proof of human courage and energy," wrote the great Norwegian explorer Roald Amundsen. "These men sailed right into the heart of the pack, which all previous polar explorers had regarded as certain death."

James Clark Ross would be rewarded with sights never before seen with human eyes, the "spectacular and theatrical" mountain ranges of what he called Victoria Land, after his youthful queen, and the smoke-plumed active volcano he would name Mount Erebus, after his ship. Most impressive, though, was the sheer ice face of the barrier—the "Perpendicular Barrier of Ice," as he labeled the Ross Ice Shelf when he wrote on his chart. The sight of the barrier stunned the seamen aboard the *Erebus* and the *Terror*. Wrote one:

> All hands when they Came on Deck to view this most rare and magnificent sight that Ever the human Eye witnessed Since the world was created actually Stood Motionless for Several Seconds before he could Speak to the man next to him.

This man, a ship's blacksmith, wished he was "an artist or a draughtsman" so he could convey the beauty of what he saw. Another, the *Erebus*'s surgeon, never left the deck for twenty-four hours, so mesmerized was he by the unbroken wall of ice.

Reading about Ross as an adolescent, I was convinced that here was the perfect calculus of the explorer. To head off into the unknown, face an insurmountable obstacle, overcome adversity, and be the first to lay eyes on unimaginable wonders.

There are a couple ways to approach stories like this. One way, the easiest way, is to get swept up in the adventure of it all. Ross faced what was considered certain death and went forward anyway? Wow! But another way is to consider what it took for Ross and people like him to accomplish what they did. What elements of courage, leadership, and foolhardiness made up that fateful decision to brave the ice pack?

So, yes, there is an element of adventure in the accounts of our own polar expeditions. None approaching that of James Clark Ross and other early explorers, of course, but ice crevasses crossed and shipwrecks survived and leopard seal attacks warded off.

But behind these stories, and I think more important than them, are leadership ideas gained by painful experience. Lessons learned by early explorers, and ones that I discovered following behind them. These ideas are the true human legacy of Antarctica. They apply to any enterprise where team building and long-term effort are natural, essential elements. They comprise a third important element of this book.

From the first, my focus on leadership has always concerned the question of how leaders go the distance. How do they sustain leadership over the course of an arduous, extended expedition? During a political administration? Throughout the history of a business? I have come to believe that leadership might be in abundant supply in this world—after all, many people will step forward and offer to lead—but sustainable leadership is a rare and vital commodity.

One reaction of listeners when I tell our story continues to surprise me. They enjoy the descriptions of a forbiddingly beautiful land, which most of them will probably never visit. But more than that, they want to know how I put together my expeditions. How did you do it? Amid the nitty-gritty details of mounting a complicated enterprise, building a team, and enlisting support, they sense the most immediate application to their own struggles, their own lives.

My most cherished hope is that the ideas of sustainable leadership, developed during treks and expeditions, can in some small way be applied to the biggest environmental test facing us today—how to keep the planet human-friendly and hospitable to life, reversing the degradations of the industrial era.

I never want to be pigeonholed and filed away dismissively in the "green" box. To me, the environment isn't a cause, it's everything. Where environmental issues meet leadership concepts, where dreams meet the effort to make them become real—in that crucial and ultimately compelling area is where I want this book to live.

The year 2041 is a deadline and a challenge. If we don't renew the treaty, if we allow mining and drilling to rip the guts out of this starkly beautiful, austere, terrifying continent, it won't just mean the failure of my insignificant, mostly ineffectual efforts. It will mean our failure as a people, as a species, to protect the planet that gives us a home.

I'll be honest with you. In the beginning, the fragile polar environment was not much on my mind. I organized my first Antarctic expedition to test myself and, as I stated, honor my heroes, Robert F. Scott, Ernest Shackleton, and Roald Amundsen. Even after I arrived at the pole on January 11, 1986, after a trek of nine hundred miles, I still thought primarily of adventure, not conservation.

That's because, as a young boy, I had fallen under the sway of bad influences: swashbucklers, sea captains, and intrepid explorers. Any plans my parents might have had for me to lead a quiet, straightforward, *normal* life were disrupted by the allure of the heroic quest, which I first discovered in a movie I watched on TV in the warmth of my own home in Britain when I was eleven years old.

Scott of the Antarctic

In 1947, Ealing Studios, the British film company, embarked upon an ambitious undertaking: to make a movie chronicling Captain Robert F. Scott's doomed expedition to the South Pole.

Ealing is officially the oldest film studio in the world, with movies being made on the same West London site since 1898. It's famous for its post–World War II comedies, such as *Kind Hearts and Coronets*. But *Scott of the Antarctic* was a different beast altogether, a polar epic of the kind not rendered on celluloid since *Nanook of the North*.

John Mills would play the lead character. He is probably best remembered today for his role as the father in *Swiss Family Robinson,* or perhaps as the father of Hayley Mills, the Disney actress. At any rate, as a father. Mills won an Academy Award playing a mute village idiot in Ireland for the film *Ryan's Daughter*. (His acceptance speech at the Oscars included the memorable line "I was speechless for a year in Ireland, and I'm speechless again now!") He made for a natural Captain Scott. His turn as a mute idiot aside, he was forever the paterfamilias, the commander, the calm and imperturbable authority figure. In *Scott of the Antarctic,* Mills stood in for Scott, and Norway locations stood in for the Antarctic.

The film is a straight-ahead hero story, a condensed chronicle of the 1911–12 expedition that cost Scott his life. The movie occupies a minor place in cinematic history primarily for Vaughan Williams's brilliant, brooding score, later worked up for the composer's *Sinfonia Antarctica*.

The BBC wheeled out *Scott of the Antarctic* dependably every year at Christmas, just as networks do with *It's a Wonderful Life* in the States. I clearly remember seeing the film for the first time on TV, surrounded by tree ornaments, ripped-apart present wrappings, and relatives gathered for the holidays. ("Tell everyone to be quiet, please!" I pleaded with my mother.)

Watching it as a transfixed eleven-year-old, I wasn't thinking about film scores or Academy Awards. After the first half hour, I wasn't much aware of watching a movie at all. I was *there*. John Mills *was* Robert Falcon Scott, and I was with him in the Antarctic, feeling the polar winds sweep across the barrier, aching in harness as I labored to pull the sledge, sorrowing as each bullet put a sledge pony out of its misery.

The arc of Scott's story struck like a cutlass at my heart. The incredibly brutal struggle to get to the South Pole. The deflating discovery that a rival, the brilliant Norwegian Roald Amundsen, had bested him by five weeks. The depths of despair in Scott's words: "Great God, this is an awful place." Then the agonizing trip back, with the defeated explorers slowly starving and freezing, finally perishing in an anomalously harsh summertime blizzard a mere eleven miles from safety.

The ground was very well prepared for eleven-year-old me to be impressed by *Scott of the Antarctic*. Like all British schoolchildren from the time he died on the barrier in March 1912 until midcentury and beyond, I was programmed to think of Captain Scott as a shining ideal of heroism and noble sacrifice. He was an icon.

Scott had been particularly useful to the British Empire in the years immediately following his death, as part of the government's attempt to convince the young men of the United Kingdom and the Commonwealth to sacrifice themselves nobly in World War I. A million of them did. Their sacrifice, in turn, further burnished Scott's standing as one of the finest English heroes of his age.

We tend to lose track of how ideals are handed down. When I was in school at age ten, I had male teachers who were perhaps a little older than average. This was because so many British men were wiped out in the Second World War. So my teacher was sixty, rather than thirty or forty. In other words, I was being educated by someone who was educated by someone who was born in the 1860s or 1870s.

The old-fashioned values came through loud and clear in Scott's story. I am trying to imagine a parallel to Scott in American life or in other cultures. The closest I can come in regard to the United States is the example of Martin Luther King. America was attempting a social revolution, a transformation whereby a single dominant race and culture was being replaced by a more inclusive, more diverse society. This was a massive social undertaking. It needed a figure around which to coalesce. It needed a myth and a hero. Rev. Dr. Martin Luther King provided that. He helped lead the way to a new kind of American society.

The comparison isn't perfect, but in much the same way, Britain in the second decade of the twentieth century was attempting a massive social undertaking, too—of a much different kind. The country needed to inspire its people for war. It, too, needed a figure to symbolize this effort, a myth and a hero. Scott served well in that respect. He became as vital in Britain's effort as King became in America's.

The late Harry Patch, born in 1898 and the last surviving U.K. veteran of World War I trench warfare, said that "Captain Scott was a great example to me." Henry Allingham, two years older than Patch and another World War I veteran, voiced similar sentiments. Patch and Allingham were both alive and cognizant at the time of Scott's South Pole expedition. They were both teenagers. The story was breaking news.

I probably first encountered Scott's story in a children's book, one of the ubiquitous Ladybird editions of small illustrated tomes that were distributed all over England. *An Adventure from History* is the subtitle of the Captain Scott number in the series (which was written by someone called L. du Garde Peach, a British author's name if there ever was one). The Ladybirds were a constant presence in post–World War II English childhood. They were usually rollicking tales, but they also quite openly served a didactic purpose. *This is how to act* was the not-quite subliminal message. *This is how heroes behave.*

Even for a child growing up a half century after Scott's death, the mythology could still hold sway. I wasn't immune to it. But I was, in fact, terribly out of step with the times. When most of my peers were swept up in a sixties mania for challenging authority, I obsessed over a naval captain long dead, an idol of the empire.

Scott of the Antarctic was released in 1948, and it did well at the box office. That the BBC still broadcast the film at all, nearly twenty years after it was first released, demonstrated an enduring appeal, if not of the movie, then of the mythology it promoted; and today it's still trotted out annually.

For me, *Scott of the Antarctic* sealed the deal. Amid the chaos of Christmas in the Swan household, I got up from in front of the telly a changed man. Or, since I was still literally in short pants, a changed boy.

There is a lot of talk nowadays about the concept of memes, of how ideas pass like viruses from person to person in a society. Well, at that annual Christmas broadcast of *Scott of the Antarctic* in Durham, England, in the year of our Lord 1967, I was infected with the virus—the virus of the dream, the virus of the hero, the virus of what is truly possible for us as human beings.

Perhaps many hundreds, even thousands of eleven-year-old boys saw *Scott of the Antarctic* around the same time I did. The meme did not infect them, at least not in the literal way it was to infect me and shape my life.

Part of the attraction was the pure lure of adventure, the call (to quote the split infinitive of another intrepid explorer) "to boldly go where no man has gone before." Captain Scott himself referred to "the fascination of making the first footmarks."

Another part of the attraction was John Mills as Robert Scott, because in his portrayal of a gruff leader I saw my own father, and even more so, my grandfather. Scott was the son of a brewer, a striver, a climber. My paternal grandfather, Tommy Swan, started out as a road laborer before seeing the future and becoming a tar-macadam mogul, a king of the road.

But part of it, the best part, was a test. What I saw in the person of Robert F. Scott was a challenge. A challenge to me, personally. I wanted to be in history. I was sick of just reading it.

How good could I be? How strong, how tough, how resourceful, how brave, how committed? How could I become a leader? How could I work as part of a team to accomplish great things? How could I become inspired and, more difficult and much more important, stay inspired?

Adventure is surely the most alluring, addictive, and intoxicating drug ever conceived. It has probably gotten more people into trouble than all illegal substances put together. "All problems of humanity come from a single cause," wrote Blaise Pascal. "Man's inability to sit still in a room." The eleven-year-old Rob Swan imagined himself as Scott and asked a naïve young man's fatal, life-changing question: *Could I do that?* I would never again be able to sit still in a room. The path that led me to Antarctica and, eventually, to 2041, began there.

Scott of the Antarctic to Beardmore Glacier. Beardmore Glacier to 2041. Inception, fruition, maturation. The progress of a dream. Connect the dots. Like many journeys and most quests, mine kicked off with a disaster that almost ended it before it began.

Irongate Wharf

———————

March 29, 1984. Seventy-two years to the day from Scott's last fateful entry in his journals. Irongate Wharf, the River Thames, London. It must be said that when my ship came in, it came in with a vengeance.

With as much fanfare as we could muster, we announced to the British public the launch of our In the Footsteps of Scott expedition. This would be the debut for me and my friends, people I had dragooned into joining me as members of an odd, inexplicable, and quite foolhardy expedition to the South Pole.

We would arrive grandly at the wharf in our official expedition vessel, formerly an oil-spill cleanup trawler named the *Cleanseas I,* now rechristened the *Southern Quest.*

Folly is always a good draw. Dozens of reporters and broadcasters gathered on the wharf, a large brick-and-concrete finger laid alongside the Thames within view of Tower Bridge crowded with sponsors, friends, family members, and polar enthusiasts, along with dozens of the merely curious. My mother, Em, was there, showing support for her youngest child.

A band of bagpipers serenaded the crowd. I had called up Major Patrick Cordingley of the Fifth Inniskilling Dragoons. Cordingley would go on (as a major general) to command the Seventh Armored Brigade, the United Kingdom's storied "Desert Rats," alongside U.S. general Norman Schwarzkopf in the Gulf War. I knew Cordingley to be interested in the Antarctic, because the Inniskilling Dragoons was the regiment of one of Scott's doomed expedition members, Lawrence "Titus" Oates. Cordingley acted as Oates's biographer.

My phone conversation with Major Cordingley went something like this:

"I need a band," I said.

"Where and when?" Cordingley replied.

His immediate willingness involved his wish to honor regiment veteran Titus Oates. With the Inniskilling bagpipers, our ship would be greeted at the wharf by about five hundred people in total.

"Not bad," I said when I heard the number, nodding to John Tolson, among those volunteers benighted enough to join me on my quest. "This needs to look good."

John didn't reply. He realized something was amiss long before I did. I saw him nervously eyeing our ship's approach to the wharf. He had conned Antarctic vessels before, and even though he was now acting as the expedition's documentary filmmaker, he had enough maritime experience to know what was going on.

For the trip up the river Thames to the wharf, the *Southern Quest* had taken on a professional pilot, a gent who in these pages shall remain nameless.

"Hadn't you better . . ." John muttered to the pilot as we approached Tower Bridge.

"Full astern," the pilot called and pulled the signal lever connected to the engine room.

That ought to do it, I thought to myself, foreseeing a smooth, majestic approach to our adoring, waiting public.

Nothing happened. The 550-ton *Southern Quest* remained on course and at speed, heading directly for the wharf.

"This is not going well," John said. He leapt out the door of the bridge and ran down to the bow. I had no idea what he was doing. He looked pretty frantic.

Maybe all horror happens in slow motion. This horror did. As I stood dumbfounded on the bridge beside the pilot, the ship bore down on the wharf like a battering ram. John Tolson hoisted himself onto the bow, wildly waving at media plebs and fear-frozen citizens. "Clear the dockside!" he yelled. "Clear the dockside!" Two or three in the crowd waved back, smiling, unhearing, not realizing that they were about to be bulldozed.

Whump! We piled big-time into the dock. The *Southern Quest* hit Irongate Wharf at a speed of two knots, which is 4.3 miles per hour. Bricks and concrete exploded into the air and rained downward. No one was hurt, but everyone was terrified. They ran like frightened chickens, emitting small mewling clucks of fear.

In the Footsteps of Scott started out not with a confident stride but with a stumble.

The river pilot had failed to realize that the *Southern Quest* was not a modern ship but an old fishing trawler, unaccustomed to the bright lights and big city. "Full astern" was an order that would be conveyed, in due course, via an ancient assembly of gears and levers, to the engines—but not before a full twenty-second lag. That twenty seconds had been long enough to send us into the wharf, scattering the crowd.

But the ship's reverse screw finally kicked in, and the *Southern Quest* withdrew from the wharf, bouncing off with a screeching groan and a cloud of concrete dust. John came back up from the bow. We looked at each other. It was too dreadful to be funny.

The *Southern Quest* pulled away from its faux pas, leaving a visible scar on the wharf, retreating as if in shame, back, back, backward . . .

"Um, hadn't you better . . ." John muttered to the river pilot once more.

The pilot yelled, "Full forward!" and engaged the signal lever. Then we crashed into the Tower Bridge.

Metal bits, this time of material dating to the mid-nineteenth century, rained on us. One of the *Southern Quest*'s masts got bent backward. The crowd on the wharf, reassembled, stared at us open-mouthed. Were these expedition dudes out to destroy the whole waterfront?

Dismay mounted in my gut like vomit. Five minutes later we were docked, this time without a hitch, and it was time for me to sprint down the gangplank and appear upbeat. Like a winner! Just don't trip on the large chunks of wharf that your bungling ship just blew sky-high! I gave a confused glance to the pilot and headed out to embarrass myself further.

The crowd was silent. Aggressively silent, I might recall. The band of the Inniskilling Dragoons had wheezed to a halt. No silence is quite as complete as when bagpipes quit playing.

What could I say? How could I possibly snatch a semblance of triumph from such an ignominious blunder?

My prepared remarks vanished from my mind. Only the abject seemed appropriate. *Sorry about that, folks, terribly sorry, I apologize,* grovel, grovel, scrape, grovel. Pasting a smile on my face, I jogged out to the gangway. Halting dramatically at its top, looking down at my shell-shocked audience, I raised my arms. A long, agonizing moment. A blush of embarrassment spread across my face.

I know there is a god, because suddenly, out of nowhere, I came up with the exact perfect thing to say.

"That, ladies and gentlemen, was a demonstration of our ice-breaking capabilities."

Well, okay, maybe not the exact perfect thing. But at least they laughed. I don't know if they bought it, but it was enough to break the mood of pure and utter folly. I strode down the gangplank and to the strobe of flashbulbs began to glad-hand my supporters. The champagne glasses had concrete dust on them.

"STARTING OUT WITH A BANG!" read one of the many headlines in the London newspapers the next day.

Disaster makes for good copy. If the *Southern Quest* had docked smoothly that day, we might not have gotten any headlines at all. The untoward incident raised the profile of the expedition, which turned out to be a very good thing for our fund-raising efforts. This cloud had its sterling lining.

Though I didn't realize it at the time, the crash would also prove to be a harbinger of much greater catastrophes to come.

Sedbergh

I grew up the youngest child in a large family. My father, Douglas, was very tough, carrying on the legacy of his father, Tommy, the young road laborer who had woken up one day and declared to himself, "I have seen the future, and it is tar macadam."

Tommy grew wealthy from tar, that sticky by-product of the petroleum industry. He became a gifted ballroom dancer but remained a tough nut to crack. Swan family legend has it that he squirreled his money away in a numbered Swiss bank account, had a terrible argument with my father just before he died, and never revealed the number. His will bequeathed my father two shotguns.

If my father passed on the Swans' tough-bastard nature, my mother, Margaret, known to all as Em, gave me something else. The dream, the imagination, the life of the mind, came from her side of the family. Her grandfather, Sir William Peat, became a founder and principal of one of the largest accounting firms in the United Kingdom. Peat, Marwick, Mitchell (now KPMG), accountants to the Queen. My cousin, Sir Michael Peat, formerly accountant to Her Majesty Queen Elizabeth, is now the right-hand man to Prince Charles.

I grew up without inherited money, but my family gave me something more valuable—a legacy of duty, discipline, honesty. They weren't alien words to me. Honor, truth, and respect. Old-fashioned values.

Uncle Johnny Ropner showed me the way in his habitual practice of planting trees. He inspired me to believe in the small ways to make a difference. "We are merely custodians," Uncle Johnny said to me. "We're going to have to give this over in better shape than we found it." Even as he said this, a spade in his hand and a burlap-balled sapling next to him, I felt the sentiment burrowing its way into my spiritual DNA.

The Peats were people of the book. The family went to Wycliffe

Church, named after John Wycliffe, the religious dissident and an early translator of the scriptures. Although I was rough-and-tumble at the prep school I entered when I was eight, Aysgarth, I had a bookish side, too.

My childlike sensitivity was in for a bumpy ride at boarding school, which represented a non-negotiable step in my upbringing. Being sent away to school was, simply put, a tradition. But tradition had devised a system that made a foot journey across the Khyber Pass look appealing.

I found it very hard. I cried a lot. I knew it broke my mother's heart for me to be away from her at such an early age. Part of me, the emotional heart of my being, gradually shut down. I bucked up. *I've been dumped into this shit, and I'm going to make it work.* That was my attitude.

Athletics were my refuge. I set myself a target to be on every school team—rugby, football (a.k.a. soccer), cricket. And I discovered an intellectual mentor, Stuart Tate, a senior master at Aysgarth and my history teacher.

Tate used his summer vacation to Italy to develop a slide show of the Colosseum that made the Roman Empire come alive for our unruly, mostly indifferent class of public school students. I was enthralled. He helped me see the bears coming up from the pit by hand-cranked elevator, the gladiators waiting, the spectators roaring for blood.

I saw *Scott of the Antarctic* at home while on holiday from Aysgarth. But I was still under the sway of Stuart Tate's vision of living history. He had switched me on to the past, and now here was a heroic slice of it that somehow seemed to speak directly to me.

Scott's body was found in 1912. The same year, the *Titanic* sank into the chill waters of the northern Atlantic. Throughout my school years, both stories, Scott and *Titanic,* were pounded upon mercilessly as propaganda to inspire young people about the nobility of sacrifice.

The male passengers of the doomed ship nobly allowed "women and children first." The British explorer nobly wrote his inspiring last words as the Antarctic blizzard howled and the last breath of life went out of him.

What was true for me had been true for generations of British children before me. John Murray's school reader of Scott's expedition journals did not go out of print until 1984.

Journalists, authors, and speech makers likewise flogged Shackleton's story mercilessly, though with a small modulation. His was a fable not so much of noble sacrifice as of leadership in extremis. Against almost impossible odds, shipwrecked and marooned, he came through without the loss of a single man under his command.

Both stories posed the same question, perfect for shaping a child's character. *Could I do what they did?* Both stories became deeply imprinted on the communal character of Britain. Scott and Shackleton. Their names were like national brands.

I carried the polar mythology with me to my secondary school, Sedbergh, in the Lake District of Cumbria. The officially sanctioned toughening process begun at Aysgarth continued there. The physical was as big a part of my education as the mental. Sports as well as academics were heavily favored. Other schools would almost rather die than play Sedbergh at rugby.

The Sedbergh School run was a ten-mile course up hills, down valleys, and across streams. We did it once a week in running season, in short pants in all weathers, snow and mud and rain, through the wilds of Cumbria. I knew every inch of that tortuous course. Later on, during my walks to both the South and North poles, I had a picture in my mind of the Sedbergh School run.

On the walk to the South Pole, we established our quota at twelve miles a day. Because of the Sedbergh run, I knew exactly what twelve miles felt like. Thousands of miles away from the Cumbrian fells, I found myself imagining *that rock* or *that tree, that hill* or *that ravine,* all hyper-realized elements of the run I had repeated so many times in school, now invoked as mental landmarks of my progress across a featureless polar plain. The strategy helped me keep myself going.

The town of Sedbergh itself helped develop the other, nonathletic side of me. It became famous as a "book town," one of the United Kingdom's small, usually rural villages that bring together a number of bookshops and other businesses based on writing, reading, and publishing (other book towns are Hay-on-Wye in Wales and Wigtown in

Scotland). At Sedbergh, I searched the bookstores for memoirs, journals, and histories of the Heroic Age of Antarctic Exploration. The period earns its capitalized status by virtue of the numerous fatalities recorded on Antarctic expeditions from 1901 to 1917: seventeen deaths in less than seventeen years.

The beginning and end dates of the Heroic Age vary from historian to historian. I'd personally kick it off all the way back in 1841, with Captain James Clark Ross's groundbreaking (ice-breaking?) voyages of discovery.

In my late teens, I read about the polar explorers whose stories would haunt me for the rest of my life. Sir John Franklin's loss while on expedition in the Arctic triggered repeated rescue efforts that killed many more men. The great Norwegian polar pioneer, Fridtjof Nansen, the first man to traverse Greenland, later went on to found an organization, the Nansen International Office for Refugees, that won the Nobel Peace Prize.

I read and reread *Scott's Men* by David Thomson. I ran through the whole pantheon, from the competing North Pole claims of Robert Peary and Frederick Cook to such relative unknowns as the Japanese explorer Nobu Shirase. I made an armchair analysis of the American explorer Robert Byrd's claims to have flown over both poles.

But I obsessed over three figures more than all the others: Captain Robert Falcon "Con" Scott, usually styled "the doomed" or "the tragic"; Roald Amundsen, his Norwegian bête noire, the coldly consummate tactician to Scott's high-wire emotional amateur; and Ernest Shackleton, Scott's other bête noire, a former expedition mate turned rival.

The heroic stories of Scott, Amundsen, and Shackleton that I found in the secondhand bookstores of Sedbergh and consumed so avidly inspired me right out of the life my father had carefully planned for me, right out of Britain, and, eventually, to both ends of the earth.

Cape Town

Timothy Leary wrote about the stages of modern life being something along the lines of "childhood, school, college, career, insurance, retirement, funeral, good-bye." I wasn't a big follower of Leary or a big swallower of LSD, the drug he so fervently promoted. But by the time I left Sedbergh I knew that I didn't want to strap myself to the wheel in precisely the manner described by Leary.

My father had other ideas. I was already getting welcome letters in the mail from his university. "We're looking forward to another Sedbergh man," one letter read, "playing and getting his blues for Oxford."

Oxford. That's what Swan men did. They went to Sedbergh and then they went to Oxford. I was a Swan man, ergo, the next step in my life would be Oxford.

Douglas Swan summoned me to his office to inform me formally of the inevitable. "You'll be starting at Oxford, Brasenose College."

A vision of a life that was not mine.

"What if I don't want to?" I said.

I remember his leather office chair creaking as my father rocked forward to face me. He looked at me over his half-moon reading glasses. My rebellion had not come out of the blue. He knew me well enough to know that I might not slip quietly into the life he had tailor-made for me.

"I don't want to do it," I said, by way of making myself clear.

"Right," he said. "Are you sure?"

I was calm. "Yes, Dad."

The theme from *Chariots of Fire* ought to be welling up right about here.

He thudded backward on his chair, then lurched forward again. "Are you absolutely sure?" The threat in his voice was clear.

"Sure," I said.

"Not only have you embarrassed the school," he said, meaning Sedbergh, "you have embarrassed the college," meaning Brasenose. "And you have embarrassed me."

I could not quite picture the five-hundred-year-old stone façades of Radcliffe Square quivering with shame over my apostasy. My father's embarrassment was another matter. Why did I resist him?

"You are now entirely on your own," he said. "I will give you nothing."

He could not very well afford to give me much. But the one thing that he could provide, a first-class education, was the one thing I turned down.

I stood up and offered him my hand. He stood and shook it curtly in the oddly formal dance of English parenting.

I turned to go, but he spoke. "If you are ever injured, my boy, I will see to it that you get first-class hospital care."

I nodded. This was all too much.

"Anywhere in the world," he added. He was well aware of my grand plans for the future, even if he did not approve of them.

That day in my father's office helped make me. But as soon as I stepped out the door, I was flooded with the conviction that I had made a bad mistake. I was on my own.

I proceeded to punish myself for it with a series of horrendous jobs. I laid pipe on the moors. I rebuilt a car. I joined a crew of Irish laborers working on a tunnel extension for the train connection to Heathrow Airport. The micks thought I was a toff-nosed little wanker, but I learned more from them about women, toughness, and drinking than I did from anyone else in my life.

All the time a refrain went through my brain. *What do I do now?* Throughout my school years, I expected that I would be at university at this point in my life. Going to Antarctica and following in the footsteps of Robert F. Scott seemed like an idea from another world.

When in doubt, try the geographical cure. I decided to leave Britain and head to South Africa. My mother saw me off at our local train depot, Darlington Station, where the first rail locomotive in history began its run.

"Rob, this is going to be tough for me as a mother for my youngest child to go off and do these things," Em said. By "these things," she meant adventures, expeditions—I was constantly talking about following Scott—and hieing off to Africa at the drop of a hat. "You'll understand when you get a little older and have children," she told me. "It's hard for a mother, so I am going to try the best I can to become your friend."

"Thanks, Mum," I said, not understanding a whit of what she was going on about.

"There might be times when I can't help it, when me being a mother will just come out, but when you do these things—not *if* but *when* you do them—I will support you all the way."

She was right. I understood none of it until I had my own child. But then I realized what a heartbreaking leap of faith it takes to entrust the wider world with your child.

I took a boat to South Africa and enrolled at the University of Cape Town. Almost immediately, within the first week, I realized that this was a total blunder. The pedagogy of the place was maybe ten or twenty years behind the style of education that I knew in Britain.

I had left home in a postadolescent blaze of "It's my life!" I wanted to make my own mistakes, and so far, that was exactly what I had done. *What do I do now?*

Crawl back to Douglas Swan, see him peer down at me, with his piercing blue eyes over those infernal half-moon glasses of his? I couldn't stand it. Absolutely not.

I had to go back home and start all over. But I couldn't bear to do it with my tail between my legs. I had already demonstrated that I had my head up my ass, so my tail probably wouldn't fit.

The bare outlines of a plan began to coalesce in my mind. I would not go home defeated. I would return in triumph after . . . biking across the continent of Africa! Yes, that's right, Douglas Swan. Cape Town to Cairo. Put that in your pipe and smoke it.

Life lessons: First, admit to yourself that you've made a wrong move and get out as soon as possible. Second, don't let a reversal prevent a bold move.

Inexplicably, though, I encountered an obstacle. It turned out I would need a little money to fund my Cape Town-to-Cairo fandango. Who would have thought? It was a detail I must have overlooked in the formulation of my grandiose plan.

So I drove a taxi in the Cape Town harborside neighborhood. I ranged from the Victoria and Alfred Waterfront south to False Bay, Table View to Simonstown to the Strand. Baden Powell Drive, named after the British founder of the Boy Scouts, became one of my main drags. Driving a taxicab would be a skill set I would find a use for later, and the sailors, druggies, and prostitutes I liveried around were entertaining enough.

As the new guy in town, I found myself at the very bottom of the food chain. I didn't get many fares, since the other drivers had more contacts and more seniority. But that changed when I picked up a drunken Japanese sailor one night, dropped him off at his destination, and later found five hundred dollars he had mistakenly left in the backseat of my cab.

At that point, five hundred dollars would have done me. I could have embarked the next morning on my cross-continent adventure. I thought, *Well, no.* The moral imprint that my parents and my grandparents left on my character asserted itself. I drove over to the Japanese boat, docked at quayside.

"One of your guys left this in my cab," I said to the captain, delivering to him the wad of lost cash.

There ensued an immense amount of bowing. The next day, the Japanese captain contacted me. "You have shown honor," he informed me in stilted but serviceable English. "From now on, we will use only you."

I made a lot of money off that single Japanese contact. In two months I cleared seven hundred dollars, enough to sponsor my trip. An ancient but sturdy postal bike would do for wheels. Now all I had to do was bike six thousand miles.

Okay, I never planned actually to pedal every one of those six thousand miles. But I pedaled enough of them to look like a weather-beaten native by the time I reached the Mediterranean. I passed through ten countries, dodged through two wars (in what was then called Rhodesia and in Sudan), and came down with a lingering attack of bilharzia, sleeping sickness, which had me virtually comatose eighteen hours a day.

Because I traveled by bike, I cut my kit down to the absolute minimum—one pair of shorts, an old pair of overalls, a couple of old T-shirts, an old sleeping bag, a sheet of plastic for a tent. Being so clearly impoverished eliminated any problem with thieves or beggars. Beggars gave *me* handouts. I got through based on total naïveté and trust.

As soon as I arrived back in England, I went to see my father. "Dad, I think I screwed up on this."

"Yes, you did, Robert," he said, looking over the half-moons.

"I am going to university," I said.

"I wish you luck," Douglas Swan said. True to his word, he offered no financial help at all.

All the driven men I have ever encountered have had father issues. Many of them passionately hate their dads. That wasn't me. I rather loved my father. But at that precise moment, the fact that I was disavowed and alone played vividly in my mind.

Durham

It is all very well to declare the steadfast principle of following through, doing what you say you are going to do. But I soon found in practice that it was no bed of roses. I had told my father that I would get on without his help. I had no idea how I would accomplish that. On my return to Britain I was just as lost as ever.

Ranked third in England behind Oxford (which I had snubbed and embarrassed, according to my father) and Cambridge (equally exalted but in the Swan household an enemy camp), Durham University happened to be located twenty-five miles from where I grew up. I drove over to Palace Green, the school's natural gathering place, and stopped the first likely student I saw on the street.

"What's the most eccentric college at Durham?" I asked him. I knew I had to find the least traditional door upon which to knock.

He didn't hesitate. "The theological college, over there."

"St. Chad's?"

"It's run by an amazing guy," he offered. "Father Fenton." (That first likely student was named Steve Moriarty, and we eventually became fast friends, training together for sports.)

I liked Durham for its cobblestone lanes, riverside topography, and the ancient Norman castle guarding it all. Looking like a Berber but with awesome, cablelike pan-African leg muscles, I presented myself at Father Fenton's office. I don't quite know what my plan was. What would help me prevail here? Charm? A good story? Sixteen-inch calves?

I tried all three.

"What happened to you?" asked Father Fenton's secretary, responding to my disheveled appearance. Her name was Bessie Bellingham, and she appeared every inch the straitlaced college gatekeeper.

"I want to come to university," I said.

"That's a problem," she said, looking me up and down. "The university term started two weeks ago."

I told her my whole story. Idolizing Captain Robert F. Scott, spurning my father's bequest, tunnel digging with Irishmen, Cape Town. We spoke for an hour. I even threw in the tale of the drunken Japanese sailor. She left me to enter the inner sanctum of Father Fenton's office. I could here the murmur of the two discussing my fate.

The gate opened. Bessie ushered me into Father Fenton's office.

I was at first confused, since there didn't seem to be anyone in the room. "Hello?"

"So you want to come to university," said a voice from the floor.

I moved around to where I could see him, long white hair and cleric's dog collar, prone on the rug behind his desk. Bad back, he explained.

"So you want to come to university," he repeated. "What subject?"

"History," I said.

He looked up at me.

On my last leg across Africa I had ridden a train across the Nubian Desert. The train was running five days late. There was no room inside the cars, and the air in there was stifling besides. The top of the train was crowded, too, but there was a gap atop one car where no one sat. So I tied my rickety postal bike on top and rode up there.

Ever since that time, the transit conductor's constant phrase, "Mind the gap," had had a different meaning for me, for I soon realized the reason for the gap atop the train. Coal smoke poured from the stacks and funneled directly back on me, blackening my clothes, my skin, and I think my brain, too. I dreamed of going to the South Pole just to cool down a bit.

I had of course scrubbed myself as best I could before I presented myself at Father Fenton's office, but under his gaze I felt intensely self-conscious, aware that I looked like something of a wild man, with coal dust still griming my skin and hair. I could only hope that a clergyman who wore his own hair long and received supplicants while flat on his back would look kindly upon my unorthodox appearance.

He did. "Bessie has decided you are coming to St. Chad's," he said. "And you're in luck," he continued. "An American chap who matriculated and then was struck by homesickness has just left us. Come by tomorrow at nine o'clock, and we'll get your paperwork started."

I was tremendously thankful and a little nonplussed. As I left his office, I turned back. "Why did you say yes?"

"I like to have some people in my college who can bring something a little bit different to the table," he said. "I understand you are good at rugby."

I shrugged modestly. "I played at Sedbergh."

"No one from St. Chad's College has played on the Durham University team for a long time," he said. "One game is all I ask."

I managed to scrape onto the Durham rugby team with Steve Moriarty, and the dog-collar brigade came down to cheer. It made for a curious spectacle, a clergy rooting section, and one that I think might have helped psych out our opponents. Through divine intervention or superior play, Durham had a winning season.

Father Fenton attended my graduation from St. Chad's-Durham in 1979. "You know, I never checked up on anything about you, on what you achieved on your exams or anything," he told me. "I wanted to prove I could still judge people."

For a long stretch after leaving his guardianship, I gave Father Fenton ample cause to doubt his own judgment.

Scott's Last Expedition

I left Durham equipped with a modest B.A. in general arts, with a concentration in ancient history. It was a handsome university, an amazing community, but as with a lot of people, what happened to me outside of classes represented my true education.

I was graced by the beautiful presence of my girlfriend, Sabine, who encouraged me to develop my intellectual side. She spoke and read Chinese and carried herself with unstudied seriousness and elegance. And here I was, this pirate by her side. I took some scrubbing up.

Like Sedbergh, Durham was a town crowded with amazing bookstores. Soon after my interview with Father Fenton, walking a quaint lane in town, I passed Bridge Books, an out-of-the-way shop near the river. Displayed in the front window I saw a hardbound, doorstop-sized, two-volume set of Captain Scott's journals, his famed account of that last doomed expedition. Original publication date: November 6, 1913. Original price: forty-two shillings. It had a deep, navy blue cloth jacket, with gilt titles on the spine and front boards. The price on the placard posted in the window beside the volumes: fifty pounds. I blanched. The sum represented pretty much the whole of my liquid assets. It would be an absurdly foolhardy purchase. "Where is human nature so weak as in the bookstore?" said the American abolitionist preacher Henry Ward Beecher, an inveterate bibliophile. I bought the books.

I had begun as a child of the movies. *Scott of the Antarctic* formed my first imaginative touchstone. But from the moment I walked out of the Durham shop with the two-volume Scott tucked under my arm, I became a person of the book.

History is written by the better prose stylist. I have read every memoir, every journal, every account of polar travel that I could find. The Scott journals stand alone. His rivals, Shackleton and Amundsen, both published their individual accounts, and they are good and worthy in their own right. But I would rank Scott's memoirs above them, for the simple reason that he gives us more of himself as a vulnerable, striving, breathing, feeling human being.

I am not sure that opening a page on a website can compete with the experience of cracking a book for the first time. All online, connected, wired in, non-book-readers thereby miss out on something incredibly precious. I recall taking Scott's journals back to my rooms at St. Chad's. I opened the volumes with ritualized reverence. Here was a real *book*, a Plato's cave of a book, the kind the pages of which you had to cut with a knife blade just to be allowed inside.

Another book quote, from another triple-named nineteenth-century American, Henry David Thoreau: "How many a man has dated a new era in his life from the reading of a book?"

Scott's Last Expedition, read the cover page of the Smith, Elder & Co. 1913 edition. "Vol. I Being the Journals of Captain R. F. Scott, R.N., C.V.O., Vol. II Being the Reports of the Journeys & the Scientific Work Undertaken by Dr. E. A. Wilson and the Surviving Members of the Expedition."

Wilson acted as chief scientific officer on the expedition. He was Scott's confidant and right hand, who died beside him in the tent on the barrier. (The line "silence deep with a breath like sleep," quoted earlier, was one of Wilson's observations about Antarctica, characteristic of the man's soulful combining of the scientific and the religious.)

I knew Wilson from the Scott film, where he is played by Harold Warrender. But in the journals Scott renders him in flesh and blood, acting always as a steadying keel to Scott's emotional rolls and dips, deeply religious where Scott was atheistic, an artist and talented drafts-man who sketched the expedition. "Wilson stands very high in the scale of human beings—" Scott wrote in a journal entry dated May 31, 1911, "how high I scarcely knew till the experience of the past few months. There is no member of our party so universally esteemed."

Via the journals I could appreciate Wilson's artwork directly. My two-volume set came illustrated (according to its title page) with "Photogravure Frontpieces, 6 Original Sketches in Photogravure by Dr. E. A. Wilson, 18 Coloured Plates (16 from Drawings by Dr. Wilson)," as well as numerous photographs from official expedition photographer Herbert G. Ponting.

Wilson's devotion to the expedition's scientific goals lent Scott the legitimacy he needed. Reaching the South Pole was merely a "game," Scott wrote. The real substance of the expedition was its scientific research. ("Science!" exclaims Scott in the journals. "The rock foundation of all effort!!")

And "Uncle Bill" Wilson was, by early-twentieth-century standards, a consummate scientist. He spent the "wintering-over" months of 1911 completing his final report on parasites to the British Grouse Disease Commission, to which he had committed while still back in England. The game birds were disappearing from the moors of northern England, and the commission wanted to find out why. It is comical but impressive to think of Wilson bent over his research in the hut on the shore of McMurdo Sound, some nine thousand miles away from the nearest live grouse, surrounded "by a halo of feathers and unraveled entrails," tracing the reason for the bird's decimation to a minute intestinal threadworm. He had given his word to the commission, so he finished his task.

> And so it is all through [Scott continues]; [Wilson] has been consulted in almost every effort which has been made towards the solution of the practical or theoretical problems of our Polar world. . . . The chief of the Scientific Staff sets an example which is more potent than any other factor in maintaining that bond of good fellowship which is the marked and beneficent characteristic of our community.

Wilson had gone with Scott on his previous exploratory expedition to Antarctica, the 1901–04 mission in the ship *Discovery*. It's very clear that Scott would have hesitated to attempt the pole without his old friend by his side.

Scott took a newly minted associate with him to the pole, too. This was Lieutenant Henry Robertson Bowers, called "Birdie" by virtue of his enormous hook nose. Scott did not want to sign him on at first. He was homely, with a modest affect, the direct opposite of the regal Wilson. But Birdie Bowers proved himself "a treasure," as Scott referred to him again and again.

An indefatigable worker, seemingly immune to frostbite, with an uncanny head for organizing supplies, the elfin Bowers provides, in his letters home to his family, a fresh, common-man perspective on Scott, sharp-eyed but overwhelmingly positive. Scott returned the sentiment.

> Little Bowers remains a marvel—he is thoroughly enjoy-
> ing himself. I leave all the provision arrangement in his
> hands, and at all times he knows exactly how we stand, or
> how each returning party should fare. It has been a com-
> plicated business to redistribute stores at various stages of
> re-organization, but not one single mistake has been made.
> In addition to the stores, he keeps the most thorough and
> conscientious meteorological records, and to this he now
> adds the duty of observer and photographer. Nothing
> comes amiss to him, and no work is too hard. It is a diffi-
> culty to get him into the tent; he seems quite oblivious to
> the cold, and he lies coiled in his bag writing and working
> long after the others are asleep.

Scott's affection for Bowers might ultimately have contributed to both their deaths, since the leader chose at the last minute to take five men to the pole instead of the planned-for four, with a resultant stretching of provisions.

The journals of Captain Scott were extremely hard won. Tryggve Gran, the Norwegian who was the ski expert on Scott's expedition, wrote of finding the tent on the barrier with Scott's dead body inside, along with those of his friends Wilson and Bowers. All three were frozen in their sleeping bags, Wilson and Bowers prone, Scott himself half raised up, left arm outstretched toward his mates, the right hugged closely to his side, the notebook journals tucked beneath it.

Out of respect (and as the Norwegian outsider on the expedition),

Gran waited outside the tent as the recovery team went in and removed personal effects. The bodies would be left as they were, in memorial, but belongings, keepsakes, and, above all, anything written would be removed and taken back to Britain.

Scott's notebooks turned out to be frozen in the clutch of his right elbow. To recover them, those expedition members inside the tent were forced to break Captain Scott's arm.

Gran reports that it sounded like "a rifle shot" when they did so.

At university, the journals became my meditation. Away from my course work, I read them endlessly. For all its professional naval technicality, the prose displayed an appealing boyishness, and in fact the journals were edited and "brushed up" for publication by Sir James Barrie, author of *Peter Pan* and Scott's close friend. Scott, dead at age forty-three, had something of Barrie's eternally youthful hero about him. I recalled Peter Pan's line "To die would be an awfully big adventure!"

Inevitably, Scott's writings led me back to the film. In my third year at St. Chad's, I rented the town's Elvet Riverside theater. Just for myself. I didn't advertise or invite other people. I got my hands on a thirty-five-millimeter print of *Scott of the Antarctic*. The Elvet was a huge place. I made a ceremony out of it. I dressed. I wore a tuxedo jacket, a starched, placketed shirt with studs, a black bat-wing silk tie. Somehow, though, my closet came up empty on tuxedo pants, so I wore a pair of khaki shorts on my bottom half. But my top half looked splendid.

I hired a projectionist to screen the film. When he got it all set up, I bribed him with a fiver. "Bugger off and have a few pints," I said. I wanted to be totally on my own. Me and the film only. I watched the movie sitting in the dark in my ridiculous outfit.

Was there anything still there? Or was *Scott of the Antarctic* merely a craze of my youth?

It did not take me long to find out. Once again, within the first thirty minutes, I was transported to Scott's Antarctic. Half-memorized phrases and arresting tableaus from the memoirs made the movie, if anything, more powerful.

For one thing, it was the first time I had seen *Scott of the Antarctic* in color. Black-and-white TVs were still the norm, and that's how I had watched it until this screening. The movie was one of the British film industry's initial big releases in Technicolor. It sounds like a joke, since you might not think to go to monochrome Antarctica for color. But the lush greens of Edward Wilson's country home in Cheltenham contrasted well with howling snowstorms of Antarctica. The delicate blues of glacial ice came through—even though what I was seeing was actually Norway, not the South Pole.

More important, I still felt the old passion for it. I was of the last generation of British schoolchildren trained for empire. The old-fashioned values of my grandfather were so deeply embedded that they would never come loose. Yes, I was out of synch with my time. Proudly so. Despite the changes raging in the world outside the Elvet, the ideal of the hero still made sense to me.

The film flickered to a close. I heard the celluloid flapping on its reel within the projectionist's booth. I remained in my seat, an audience of one in a darkened theater.

It was a commitment moment. *I am going to do this. It doesn't matter how long it will take. I am going to follow Scott to the South Pole.*

An answering voice chimed in, *You're an idiot,* but I chose to ignore it.

Cervantes: "Folly is the one thing that makes life worth living, yet no man counts himself a fool."

I left university with a total focus. I had talked the talk for a long time. It was time to walk the white walk.

Norton Versus Kawasaki

The Scott journals were one gift Durham gave me, and great friends were another. Those were the days—the late 1970s—when the British motorcycle industry first felt the assault of Japanese imports. Dyed-in-the-wool national-pride traditionalists stuck to Nortons, Triumphs, and BSAs, sneering at the influx of Kawasakis and Suzukis.

I blew into the parking lot of the Durham University student union one afternoon on my 900-cc Kawasaki "rice burner." Pulling up beside me on a British Norton at the same time, blowing oil, another rider, a frizzy-haired beanpole of a bloke, gave me and my Kawasaki the once-over.

"Do you carry oil with you?" I asked him, responding to his scornful stare. Nortons were notoriously loose in the gaskets and tended to spit and smoke.

He wasn't offended. He smiled. "Peter Malcolm," he said, offering his hand.

It was, as Bogie says in *Casablanca,* the beginning of a beautiful friendship. Which did not start out that well. We swaggered into the student union together, and within the first two minutes of conversation I made him an offer that he probably should have refused. "Do you want a job?" I asked him.

I had a logging and tree-surgery business on the side in those days. My father had kept his promise not to support me. I had to make my way on my own. The morning after I met him, Pete Malcolm and I had picked up a chain saw at my mum's. Ten minutes after that, we were up to our knees in the icy autumn currents of the River Wear. These were destined not to be the last freezing waters into which I would lead Pete.

Removing a troublesome elm that had fallen across a home owner's property, I gunned the saw, lopping off a section of trunk. The log rolled across my foot. Pete helped me push it off and we carried on— more sections cut, more branches lopped, our feet numbed by the cold.

"What the hell is that?" Pete shouted at me over the buzz of the saw.

The Wear ran blood-red in a plume downstream from my leg. The log that rolled on me had done more damage than I had thought, the wound unfelt because of the freezing water.

Not for the last time, I was glad I had Pete with me. He got me back to the work van that I used on my jobs. The ancient vehicle started not with an ignition key but with a hand crank. I still remember bleeding inside the van, watching Pete's face turn various shades of purple as he tried to crank-start the engine.

We finally got to the hospital. Pete bonded with my mother as the doctors tended to the gash in my leg. He came to the recovery room to say good-bye.

"Peter," I said to my friend of two hours, "my bike is still parked back at the pub."

No problem, Pete replied. I'll pick it up for you.

This was the era before cell phones. An hour later a nurse wheeled a telephone into my hospital room on a little cart. When I answered, Peter Malcolm's voice came on the phone. "Rob, I've crashed your bike."

Unaccustomed to the Kawasaki's 900-cc acceleration capabilities, Pete had spun the motorcycle out on the first corner he attempted. He was unhurt, but he'd totaled my rice burner.

So, the first four hours of our friendship: a miserable job in a freezing river, a sliced-open leg, a visit to the emergency room, a motorcycle crash. Maybe we should have turned back right then—shaken hands, wished each other well, but concluded that perhaps the gods of luck would not look kindly upon our continued collaboration.

Like that scene in *The Wizard of Oz* when Dorothy and her pals venture into the dark woods, heading toward the witch's castle. They encounter a sign—I'D TURN BACK IF I WERE YOU. Just before they get ganged by a troupe of flying monkeys.

But we didn't turn back. Instead, we joined together in a dream of Antarctica.

Five Million

"If you are going to do something as stupid as going to the South Pole, Robert, you'd better get a plan." The speaker was Sir Geoffrey Gilbertson, a friend of my mother's, to whom I went for advice. He spoke to me in his office, from a wheelchair, where a bout of polio had put him.

Some years earlier, Sir Geoffrey had married Dot, my mother's dear childhood friend. They were frequent guests in our house. I admired him as a tough, no-nonsense character. He had been a top oarsman in school and a tank commander in World War II, but after the war he had been one of the last people in Britain stricken by polio. He was knighted in recognition of his efforts to increase public access for the disabled.

He wasn't the type of man whose advice I could easily ignore, and Sir Geoffrey wanted me to have a plan.

A plan! What a novel idea! I was such a naïf that even as I was swept up in my grand idea to walk to the South Pole, I hadn't developed a coherent approach as to how to go about it. I thought it would just *happen*.

Going to Sir Geoffrey represented an early instance of what would eventually become one of my standard operating procedures. Again and again over the course of my life, I have had to remind myself not to be embarrassed to ask for help. Or, even if I *was* embarrassed, to still ask for help. A simple rule, but one that a surprising number of people ignore.

"What about a budget?" Sir Geoffrey asked.

I looked at him helplessly.

"You are going to ask people for a great deal of money," he said. "They will want to see where it's going to go."

After I graduated from Durham, my quest to follow Scott's path across the Antarctic began to crank into high gear. Actually, to follow the metaphor exactly, there first came the gnashing of gears and the spinning of wheels. I became tiresome. I reached out to anybody or anyone who I thought could give me some clue as to how to go about doing this expedition I wanted to do, which seemed more and more impossible the deeper I got into it.

I talked to my family. "Rob, if you don't do it now, you will never do it," said Jonnie, my oldest brother, sixteen years my senior. Jonnie worked as a university lecturer in taxes and accounting.

I thought I could detect a gleam of our old childhood energy in Jonnie's advice to me. If you are going to go off on a wild-ass tear, he seemed to be saying, then the mid-twenties, the age you are now, is the time of your life to do it.

Okay. All right. Do it now. But do what *now?*

I followed Sir Geoffrey's advice. I made a budget. After totaling supplies, transport, contingency funds, and random expenses, I came up with a number. A ballpark figure, as they say in America.

I stared at it. Staring at it did not make it diminish.

Five million dollars.

I was twenty-five years old. Who ever heard of a twenty-five-year-old raising five million dollars?

I sent Sir Geoffrey a note informing him of the number I'd come up with. He sent the note back with the word "Forward!" scrawled across it.

I told Pete Malcolm, too. He had joined the Royal Navy, but it was becoming another standard policy of mine always to update my support network regarding any developments. Just occasional bulletins, nothing constant, but it kept everyone in the mix. I knew I would need all the help I could get.

I had a budget, featuring an absurd number that seemed to push the dream farther away rather than closer. I groped for a next step. I realized I had to accomplish something concrete, something not on paper.

Well, I told myself, *you are living the pages of Scott's journals.* I read them so often that passages were imprinted in my brain. I knew everything in them, every man, all the events, each accomplishment and setback. Mentally, I went back through *Scott's Last Expedition.* How

did Scott start? He embarked on his expedition from the Thames docklands. So that's where I went.

Shed 14, West India Dock, was where Scott loaded his ship. The
place had become part of Canary Wharf, a stretch of warehouses and
docks along the riverfront in Tower Hamlets. I rented space in one of
the warehouses nearby, on Wapping's Metropolitan Wharf, for one
hundred dollars per month. At that time, the area was run-down and
decayed. Nowadays, in upwardly mobile and gentrified Canary Wharf,
the same space might cost ten thousand per month. I paid rent for the
privilege of occupying a dank, musty, unheated, and quite rat-infested
warehouse, floored in stained concrete. On summer days, when the sun
heated the bricks, we could still smell odors of spices and tobacco from
the glory years of empire. But more often the heady aroma of history
was masked by the pungent smell of the nearby river Thames.

It didn't matter. The journey of a thousand miles (or ten thousand)
begins with a single step. Our own warehouse. We had begun.

By "we" and "our," I mean myself and the only two recruits I had
managed to entice into the new digs. William Fenton—no relation to
Father Fenton—was another of my classmates at Durham. Will was an
outstanding oarsman who actually left university to come down to the
Isle of Dogs Docklands to work alongside me.

Will Fenton acted as the effort's word man, creator of the elaborate
news releases that at first no one would bother to read and the
brochures no one would accept. But he became absolutely indispensable in eventually getting the expedition off the ground.

The other recruit was my girlfriend, the beautiful Rebecca Ward, former Playboy Bunny, glamorous girl-about-town, reduced through her
association with me to answering the office phone and licking envelopes
for our fund-raising campaigns. She used to wrap herself in multiple
sweaters to endure the dockland warehouse's freezing temperatures.

We had a brass nameplate made up for outside our front door that
read SCOTT ANTARCTIC EXPEDITION. We were very proud. We wrote
letters. Rebecca mailed them. Will contacted the media.

Almost without warning . . . nothing happened.

The Secret of the Pole

Early 1980. While all this nothing was happening I somehow had to find a way to live. I drove a taxicab, putting to use the hack chops I first developed in Cape Town. I kept an eye out for drunken Japanese sailors.

Dutch elm disease had hit Britain, and with my brother Roderick I did tree surgery and removal. He represented the true brains of the Swan-Usher clan, at Gordonstoun School with Prince Charles, graduating from St. Andrew's with double first-class honors in moral philosophy and political economy. He built his tree business into one of the United Kingdom's largest.

Many of the diseased trees were in back gardens with no access except through the house itself. Roderick and I would chainsaw the tree into luggable lengths, then haul it outside, piece by piece, through the client's downstairs parlor. *Pardon me, ladies and gents. Coming through.* Hauling four-foot sections of elm turned out to be suitable physical training for the rigors ahead of me.

Mostly though, I butted my head against the wall. I wasn't even getting doors slammed in my face, because that would imply that there were doors that opened to me even briefly. In those days the idea of sponsorship had not developed into the widely accepted, finely honed practice that it is today. There were no sports stars changing ball caps mid-interview in order to secure both brand names during their product-placement time.

Our sole success: through the good contacts of my schoolmate Frank, a brewing company called Vaux donated three thousand cans of beer to the cause. Fitting, I thought. Scott was the son of a brewer and suffered lifelong feelings of social inferiority because of it.

The beer, of course, was long gone before our stillborn expedition ever saw a glimmer of Antarctica. These were the days when I took comfort from the story, probably apocryphal, about the founder of Kentucky Fried Chicken proposing his idea in three thousand meetings, getting turned down time after time, before he found success. *I've done nine hundred,* I told myself. *Now I've only got two thousand to go.*

"What are we doing wrong?" I asked Will Fenton.

"We need sponsors," he said.

I resisted a sarcastic "well, duh" response. But his comment triggered a thought. What sponsors had supported Scott? Maybe the same firms were still around. Brilliant! We could ask for funding from them in the spirit of tradition, citing their aid to the original expedition.

"Scott must have been supported by somebody," I said, excited. "It can't all have been the navy and the Royal Geographical Society."

But as I pored through my bible, the Scott journals, I found no called-out lists of sponsors. A few hints and references but nothing concrete. Who had backed him? It was a mystery.

All along I had been firing off letters to Scott's son, Sir Peter Scott. He barely knew his father, who froze to death on the barrier when Peter was three. *Peter Pan*'s J. M. Barrie acted as his godfather. Robert F. Scott's last letter to his wife, Kathleen, Peter's mother and a charismatic sculptor, charged her to "try and make the boy interested in natural history if you can."

In this, Kathleen Scott succeeded admirably. Peter grew up to become one of the world's foremost protectors of wildlife and the environment. An Olympic-medalist yachtsman, pioneer glider pilot, ornithologist, accomplished painter, and renowned conservationist, he was a founder of the World Wildlife Fund and designed the organization's instantly recognizable panda logo.

We badly needed Sir Peter Scott's approval and imprimatur. A positive word from him would instantly transform our little warehoused enterprise into a going concern. The lack of his blessing, on the other hand, represented an awkward gap in the ramparts of legitimacy that we were attempting to erect.

But Peter Scott did not respond to the importunate letters of a nobody named Robert Swan who wanted to follow in his great father's footsteps. One accommodation I did manage to obtain allowed me to examine Scott's papers, housed in the Scott Polar Research Institute at Cambridge.

QUAESIVIT ARCANA POLI VIDET DEI, read the inscription on the northern façade of the buff stone institute building on Lensfield Road. *He sought the secret of the pole but found the hidden face of God.* I pondered the words without being entirely sure I wanted to seek or find either.

The building contained a library and small museum, as well as offices for such august bodies as the International Whaling Commission, the International Glaciological Society, and the Scientific Committee on Antarctic Research. In the garden, a heroic sculpture by Kathleen Scott (her model was the younger brother of Lawrence of Arabia), and in the entablature, a bust of her husband, also by Lady Scott. The place was haunted. I felt out of my league, a spy in the house of history.

But in the papers kept in the institute's archives, I found what proved to be the key to mounting my expedition.

Onionskin

As Captain Scott's banker," I told the lower-level executive from Barclays Bank, "your firm gave invaluable support to the original expedition. I am confident that you'll see fit to support a contemporary expedition to honor your original arrangement with Captain Scott."

"Yes, I know," said the executive. "I mean, I see by your letter that you believe that Barclays acted as Captain Scott's financial institution. But my research shows that is simply not the case." The man from Barclays gave me a thin smile.

"I guess I should have been a little more clear," I said. "Captain Scott's bank, Cox & Biddulph, is no more, having been acquired in 1921 by"—I pretended to look at my notes, making him wait for it—"Barclays Bank."

It was my turn to return the thin smile. And it was true. What I found in the Scott Polar Research Institute archives was a trove of letters on the almost-transparent onionskin stationery of the day, cracked and faded but still readable. I called them "we remain your obedient servant" letters, since that's how Scott signed off.

Letters to biscuit companies. Letters to mackinaw mills. To manufacturers of patent stoves. Tinned beef. Marmalade. Coal. Rubberized footwear. Letters on onionskin stationery to firms that sold onionskin stationery. And quite a few communications between Scott and his bankers at Cox & Biddulph. We remain your obedient servant.

Mr. Lower-Level Barclays Bank Executive passed me up to a Mr. Middle-Level Executive, who passed me up to Don Pratt, the most amazing—and the most patient—bank manager in the history of the planet.

Under Don Pratt's guidance, and with my willingness to mortgage my smile, my future, and any loose assets that came my way, Barclays Bank helped underwrite my Scott expedition. The funding started coming in. Slowly, but it came.

What I learned is that to raise money you have to show people a big picture. You have to sketch your dream in broad strokes. And you have to oversell. If you are trying to raise $100, you will spend $5 dollars to raise that $100, so your real budget is $105.

In searching for backers, I regularly encountered a version of the seminal question of "Why?" Why walk to the pole? Why go to Antarctica at all?

I recalled a scene in *Scott of the Antarctic,* when Scott was trying to raise money for his expedition. He addresses a group of good British merchants, who listen to his descriptions of beautifully desolate landscapes and heroic adventures with sober, bored expressions.

Finally, one of the merchants speaks up. "What's the prospect of trade between this city and the South Pole? Is there any coal down there? Anything that I can buy and sell?"

That was the only why that some people could accept—the possibility of profit. It was clear-cut, simple, without messy subjective judgments about beauty and heroism.

George Mallory might very well have been the first man to climb Mount Everest. He died there, and his body remained undiscovered until 1999. Mallory answered the why of scaling Mount Everest with the best-known words in mountaineering: "Because it's there."

I couldn't do much better than that, but the truth is that the why of the matter seemed to change constantly over the course of organizing the expedition. As soon as I tried to pin it down in my own mind, it slipped out of my grasp. I had a hundred reasons, and I had none. Better to concentrate on the here and now, and the task at hand, and let the larger questions answer themselves. I threw myself into the work.

We had to find ways of showing the value of our fund-raising efforts. A donation of ski equipment, for example, could be shown on the books as worth a certain amount. Totaling all the in-kind donations, barters, and trades helped us to portray In the Footsteps of Scott as increasingly solvent. It lent a momentum to the expedition.

It's difficult to understand the primitive nature of office technology back then when compared with all the Wi-Fi bells and whistles we have today. It was the Stone Age. We had no computers, no fax, no cell phones. I remember traveling around London with a box of coins for the pay phone. Actually, the first cell phone was just coming into use around that time. After our effort gained a little momentum, In the Footsteps of Scott leased one. It was the size of a small suitcase, portable only in the broadest possible sense of the word. But there was no e-mail, no Internet. When we wanted to research a fact or a piece of history, we had to damn well send someone down to the British Library to do it, and it usually consumed the better part of a workweek.

Technology was one element. Money was another. But by far the most important was people. I was having as much difficulty finding the right people to help me go to Antarctica as I was finding the funding to get me there.

I still had not connected with the leading names associated with polar exploration, luminaries such as Sir Peter Scott, Lord Shackleton (son of the explorer), or Sir Vivian Fuchs, best known for the first overland crossing of the Antarctic Continent. In the eyes of the British polar establishment (yes, there is such a thing) my expedition did not exist.

In the classic catch-22 of the polar establishment, in order to be able to go to Antarctica you had to have gone to Antarctica. This was long before the days of Antarctic tourism and boatloads of well-fed, geared-up adventure travelers disembarking by the hundreds to snap pictures of penguins. There was no such thing as private expeditions to Antarctica. The prevailing wisdom was that if you weren't a scientist doing research, you weren't going.

But I had to go. To give the expedition any heft at all, to be taken seriously myself, I had to get to Antarctica. I knew it was the only way that I could find the people to help me on the trek. I could not do it alone.

We nursed the conviction that the only way to do this properly would be to buy a ship. Do it the way Scott had done it, by landing the expedition and wintering over. It was a massive undertaking, and we tried to approach it in other ways, but in my heart of hearts I knew that would be the way to go. For such a complex task, the expedition needed to enlist experienced Antarctic guides.

I tried everything. I attempted the geographical cure first, borrowing some money from my brother Tommy and flying to New Zealand. The Kiwis always had a strong historical association with Scott—"a great mass of people" and a huge fanfare marked both his departures from Lyttelton Harbor on his Antarctic expeditions (first on the 1901 *Discovery* and then, in November 1910, on the *Terra Nova*).

Tommy's money didn't spread much beyond airfare. I distinctly recall being so broke that I slept under a tree. In the morning I brushed myself off the best I could and kept an appointment at the offices of the New Zealand Polar Division, Department of Scientific and Industrial Research.

I presented my case to Bob Thompson, director. Scott and New Zealand, a match made in history, the heroic tradition, win one for the Commonwealth, rah, rah, et cetera.

Thompson was not much impressed. "Well, you're not a mountaineer," he said, stating the obvious. "And you're not a scientist." Sounding vaguely accusatory.

What could I say? I indeed was not either of those things.

"We'll keep you on the list," he said. Meaning, on the contrary, that he would definitely not keep me on any list anywhere among the myriad lists in his near vicinity.

The flight back to the U.K. depressed me mightily. Another closed door. I thought, *I will never do this.*

But I refused to quit. I assembled all my letters, all my references, all my charm, and went to an appointment at the British Antarctic Survey (BAS) in Cambridge, on the opposite side of the university from the Scott Polar Research Institute.

There I got the same response, but a more sympathetic ear, from Dave Fletcher, the head of the BAS's field operations.

"Look, Rob, you'll never get a job with us," he said. You're not a scientist and not a mountaineer, he told me.

Yes, sir, I already knew that. A Kiwi chap just pointed out the same fact to me. "I need to get to Antarctica," I said. "Isn't there anything I can do?"

"The only position would be a BGA," he said.

"What's that?"

"Base general assistant."

"A slave. A gofer."

He nodded. "But unfortunately, you don't qualify." Even BGA's needed credentials in Antarctica.

"What do I need to do?"

"Well, you look strong enough, and fit," he said. My arms were like cords from lugging lengths of diseased elm. Then he proceeded to tick off additional requirements. Experience as a mechanic on two-stroke engines. Paramedic training. Maps, photography, radio communications, skiing. "And you'll need expedition experience."

"To be able to go on an expedition I need to have gone on an expedition?"

He shrugged. "That's the way it is."

"Right," I said. "I'll be back."

Incrementally, the momentum of my dream had begun to build. Support had begun to trickle in. Not five million dollars' worth, but hundreds of thousands of dollars nonetheless. I had gathered around me a team of great, committed people: Pete Malcolm, Will Fenton, and Rebecca. From Dave Fletcher of the BAS I knew specifically what my next steps had to be.

But the expedition continued to lack the requisite endorsements. Peter Scott still declined to weigh in for or against. Help from that quarter would come from an unlikely ally.

Scott and Amundsen

At age fifty-two, the British author Roland Huntford reached a turning point in his literary ambitions. He had written a right-wing screed against Swedish socialism and a historical fantasy novel in which Christopher Columbus first journeyed to the New World in 1472, with a company of Norwegian sailors.

But in 1979 Huntford published an account of polar exploration called *Scott and Amundsen*—later adapted as a TV miniseries and reprinted under the title *The Last Place on Earth*. The book caused an uproar, although in all fairness one hesitates to call it a work of nonfiction. It is rather a quondam novelist's exercise in myth busting and pedestal toppling, a hatchet job almost comical in its extreme bias.

Huntford's *Scott and Amundsen* took as its target the iconic figure of Robert F. Scott, the tragic national hero under whose spell I had fallen when I was eleven. Just as the man on the television screen had little relation to the flesh-and-blood Captain Scott—he was played by John Mills, an actor, after all, and was for me inflated all out of proportion by a young boy's fecund imagination—so Roland Huntford's feet-of-clay Scott bore only glancing relation to reality.

In the decades since his death, the British national imagination had worked overtime to lionize Scott. His was perhaps the best-known death this side of Christ's. And it was not only the Brits engaged in this efforts: The Soviets added Captain Scott to their pantheon of heroes, intent as they were to promote sacrifice unto death on the part of the Russian populace.

Huntford clearly relished the opportunity to pop the gaseous bubble of Scott's popularity. To be sure, years of memorializing led to distortions and misperceptions. The great explorer's reputation had congealed into caricature. A corrective was due. But if the pendulum of hero worship and adulation had swung too far one way, Huntford became overzealous in pushing it back the other way.

The Scott in the pages of *Scott and Amundsen* was "weak, incompetent and stupid," as well as morbid, insecure, sentimental, and amateurish. Huntford wrote of Scott's impatience, lack of insight, refusal to accept criticism, failure in leadership, and bouts of depression. Scott's rival, the Norwegian Roald Amundsen, was careful, prepared, thoroughgoing, professional, the consummate polar explorer.

The bare facts remain identical in every account of the Heroic Age of Antarctic Exploration. Both Scott and Amundsen went to the South Pole. Amundsen made it there first. Scott died on the return journey.

British commentators fashioned one sort of narrative out of these facts. Scott became the tragic hero, Amundsen the upstart afterthought. More than that, Amundsen was blamed for Scott's death. The British hero died not of starvation or exposure but of a broken heart, having lost the race to the pole. In the public mind, Scott's death erased Amundsen's achievement. In his book *My Life as an Explorer,* an exasperated Amundsen complained that British schoolchildren were being taught that Captain Scott discovered the South Pole. The British were sore losers, he sniffed.

Huntford used the same set of facts to develop a very different but equally skewed revisionist narrative. Scott was ill prepared for the expedition. His many mistakes—chief among them, his failure to use sled dogs—doomed his expedition to failure and him and his companions to death.

Amundsen depended on reason, experience, and exhaustive preparation to win the celebrated race to the pole. He humbly submitted to learning from indigenous peoples how to survive in the hostile polar environment. The arrogant Scott declined to learn from anybody, foolishly depending on his gung-ho Royal Navy how-to.

Amundsen triumphed. Scott died.

I read Huntford's *Scott and Amundsen* when it first came out. By that time, I had evolved out of my schoolboy hero worship simply by virtue of knowing the details of the story. I had studied more books, sifted through more journals and diaries. But even though I had put away childish things, I remained impressed by the fablelike possibilities in the story.

The major Antarctic expeditions at the dawn of the twentieth century are easy to name:

- Scott's *Discovery* expedition of 1901–4, on which Shackleton acted as a junior officer
- Shackleton's *Nimrod* expedition of 1907–9, when he stopped and turned back ninety-seven miles short of the South Pole
- Scott's ill-fated 1910–12 *Terra Nova* expedition, the one our own expedition sought to honor, and which was simultaneous with
- Amundsen's successful South Pole expedition in the *Fram* ("Forward")
- Shackleton's disastrous but ultimately heroic *Endurance* expedition of 1914–16

It was difficult *not* to imagine the figures of Scott, Shackleton, and Amundsen as symbols, players who strutted and fretted their hour upon a much larger stage than just the South Polar continent. History made them less flesh-and-blood men than archetypes: Scott the inwardly troubled naval hero, Amundsen the doggedly efficient Norwegian outdoorsman, Shackleton the hail-fellow-well-met who turned fierce lion in the face of catastrophe.

The three of them sometimes appeared as human chess pieces engaged in a great game. Scott versus Shackleton, Amundsen versus Scott, all of them against the raging natural elements of the South Pole.

In *Scott and Amundsen,* Huntford played the same game, but he cheated a bit, too. Nowadays we would say he was engaged in spin. He was selective in his choice of facts and either ignored or was ignorant of important parts of the story. He showed himself to be an inveterate second-guesser, an unappealing armchair explorer.

Huntford's main mistake was that he was comparing apples and oranges. Amundsen's stripped-down expedition concentrated solely on attaining the pole. Scott's expedition focused on science. The South Pole was an element of the effort, but it was not the only element. Research comprised the raison d'être. The recovery team pulled thirty-five pounds of geological samples from Scott's death tent, collected on the escarpment of the Queen Alexandra mountains. Thirty-five pounds! Weight that they could ill afford to haul in that last failed march toward a supply depot. Even while dying from cold and starvation, Scott, Bowers, and Wilson refused to abandon their scientific aims, which Huntford airily and incorrectly dismisses as minor and negligible.

In the years since Huntford's book was published, new research indicates that Scott was done in by a spate of unseasonably bad weather, outside all statistically probable norms. In her book *The Coldest March*, Susan Solomon demonstrates that many of Huntford's assumptions about weather were quite simply wrong.

There is something unseemly about lobbing potshots at a far-off figure, separated in time and distance from the critic, who enjoys the comfortable perspective of hindsight and the security of a well-heated study. Over the course of five hundred brutally slanted pages, Huntford assassinates Scott's character, thrust by bloody thrust. The book recasts Scott the national hero as Scott the inept bungler. The man's reputation has never been the same since.

Out of the blue one day a message reached our offices in the freezing warehouse on the Thames. *Sir Peter Scott would like to see you.* This was the man whom I had been entreating forever, desperately trying to get his approval. I had just about given up.

Mentally hyperventilating, my hair plastered down within an inch of its ragged life, I met Scott in the office at his home in Gloucestershire. His own superb watercolors of birds adorned the walls.

I must convince him, I told myself. So I laid out my whole story, much as I had to Father Fenton's secretary.

But Peter Scott wasn't Bessie Bellingham. I was met with silence instead of encouraging feedback. He sat immobile behind his desk.

I chattered on. No response. I could not figure it out. Did he like me? Hate me? Was he deaf? Asleep?

After about fifteen minutes of my monologue, Sir Peter Scott put up his hand. I stopped talking. A long pause.

"What you are trying to say is that you are the right man to honor my father in this way," he said.

Yes!

"I believe I agree," he said.

Relief whooshed into me.

"This writer, this book," he said, a look of distaste on his face.

I didn't have to ask which writer, or which book. Roland Huntford. I suddenly realized why Peter Scott was assenting to support my
expedition.

"I'd like to give the blighter a bit of a hard time," Sir Peter Scott
said quietly. "I'd like to show him that my father still has a lot of support in this country."

Yes, I could use Peter Scott's name as a patron of the expedition.
That moment might have been the first time I actually believed that my
dream might come true. And I could thank the mean-spirited literary
assassin Roland Huntford for sending me on my way.

I learned later that one whimsical reason Peter Scott had agreed to
see me was that the swan was his favorite bird.

He had one request. We had been calling our effort the Scott
Antarctica Expedition, and he rightly pointed out that no one named
Scott would actually be going to the pole. With Sir Peter's blessing, we
rechristened ourselves In the Footsteps of Scott. We were one footstep
closer to getting started.

The Royal Navy Mount Kenya Expedition

W e have to do an expedition," I said to Pete, now Sublieutenant Peter Malcolm of the Royal Navy, a helicopter pilot. I had kept Peter apprised of every faltering step I took to fulfill my dream. As I said, I learned that it is important to do this. To keep your allies firmly in your alliance, be careful to keep them in the loop.

One doesn't just "do" an expedition. Before mounting such an expensive and extensive effort, it's customary to have specific aims, preferably scientific.

Our impulse was more along the lines of Judy Garland and Mickey Rooney: "Let's put on a show!"

But in order to get In the Footsteps of Scott going, I had to be able to say that I had been to Antarctica on an expedition. And in order to sign on with the British Antarctica Survey as the lowest lackey, I had to have experience on an expedition—to somewhere, anywhere.

Thus was born the Royal Navy Mount Kenya Expedition of 1980. Pete Malcolm named me the expedition's "ice expert," simply by virtue of my experience pedaling past the eponymous mountain on my trans-African bike trip.

The RNMKE was in no way a sham. We did indeed go to Kenya. We climbed the bloody mountain.

As it turned out, we didn't summit. A good lesson to learn—when to turn back. (Some of my expedition mates over the years would say I did not learn that lesson well enough.)

I am not entirely sure that Pete's naval superiors knew what he was up to, Mount Kenya being three hundred miles from the nearest ocean. But I got my official expedition testimonial on the letterhead of the Royal Navy.

I duly presented myself once again to Dave Fletcher at the British

Antarctic Survey. I had spent a couple of months working at a garage that specialized in two-stroke engines. I had brushed up on my map and radio skills. I was certified in first aid as a paramedic. And I had expedition experience. "I'm back," I said.

Fletcher heaved a deep sigh. "I rather thought you might be," he said.

So in 1981, I set foot on Antarctica for the first time with the BAS. Rothera Research Station, Rothera Point, Adelaide Island, off the west coast of the Antarctic Peninsula—not continental Antarctica, at least not yet.

Base-camp general assistant Robert Swan, at your service. A general assistant basically does whatever anyone tells him to do. Officially, I was in support of scientific research. I kept the members of the expedition who were more important than me—such as the sled dogs—supplied with food. I moved boxes.

The odd thing is that I didn't have much free time to take in the place itself. I was too busy humping crates of dog food and drums of diesel fuel to stop and smell the flowers. Of which there were none.

Rothera. A collection of low, shedlike buildings huddled beneath an airplane runway—actually not a real runway at all but what was called a "skiway," where the Twin Otter ski planes flew in from the Falklands. Fuchs House, Bransfield House, and the Chippy Shop were the largest of the buildings that dotted the site, with other, smaller sheds grouped closely around them. It was a research base in the strictest sense: Most of the scientists departed from there on research missions to other locations, using Polaris snowmobiles for transport.

If Antarctica is shaped like a fist, it has an eight-hundred-mile-long thumb sticking out as though hitchhiking (Get me the hell out of here!). The thumb is the Antarctic Peninsula. Adelaide Island, named for the last Hanover queen of Britain, is lodged along the radial border of that thumb. At Rothera we were perched on a rock, a short distance from an ice piedmont on which the Otters landed and just below Reptile Ridge (named for its spiny ridgeline, and a geological reminder that the continent was once submerged beneath a tropical sea). Across the strait rose the mountains of the peninsula, pyramids of black stone, veined with ice. If I ever did stop humping supplies, it was to gaze across the water to those mountains, beautiful in their desolation.

What was happening to me is what all visitors to the bottom of the world experience. The Antarctica of my imagination—the place I had known from *Scott of the Antarctic* and books and photographs—was being replaced by the Antarctica of reality.

Mary Shelley, in *Frankenstein,* wrote about the Arctic without ever having been there. Charlotte Brontë in *Jane Eyre* did, too—she called the far north "the death-white regions." The polar dream floats in the mind of the world as it did in my own mind, long before I ever made it there.

Antarctica is one of the rare places in the world that lives up to its mystique. In quite a selfish sense, it came through for me. Every Fid at Rothera (the term *Fid* is customarily used for Brit Antarctic personnel) helped me on my quest. I met the people there who would make In the Footsteps of Scott a reality.

Roger Mear, one of the world's premier mountaineers, had scaled the face of Eiger in winter. Roger would introduce me to Gareth Wood, an organizational genius. John Tolson had experience in polar waters. Graham Phippen was first mate on the *Bransfield,* the ship that brought us out of Rothera, on which I also met an expedition doctor named Mike Stroud.

Without fully realizing it at the time, I had assembled my team for In the Footsteps of Scott. All true leaders know that putting together a team is a crucial step in any successful endeavor. It may be the most vital process of all. While I was at Rothera, I didn't realize the real worth of the people I had met. They would prove themselves in the crucible of the expedition.

I wish I could say the selection process was somehow guided by my brilliant insight into the human character. It was not. I felt my way in the dark. Somehow, by a stroke of luck or providence, I blundered into recruiting teammates who were perfect for the task at hand.

This has happened to me so often in my life that I want to step back and consider it for a moment. Roger, Gareth, Graham, John, Mike. These people, alongside Pete Malcolm, Will Fenton, and a few others, are the ones to whom I can always turn. Not to put too fine a point on it, they are the ones who pull my ass from the fire.

I have a private, pet name for them: the Immortals. I call them that for the way they will leave my life for a while and then return, as if re-born into it, years later. They are my true friends. But more than that, the Immortals seem to be cut of superior cloth when compared with other humans—more dedicated, more capable, more intense. The high-est honor I could imagine would be to earn a place on someone else's list as an Immortal. That's what I'm gunning for in my relationships.

At Rothera, I found myself among people who would go on to transform my life.

Antarctica. I had cleaned hospital emergency rooms to get there. I had driven a taxi and humped diseased trees and worked as a barman. To accomplish this oversized dream of mine, I had to demonstrate com-mitment.

This is the case whenever you ask anyone else to join you in a dream. If you waver, if you prevaricate, it's lost. I was convinced that the reason I could attract great people, my Immortals, is that I had demonstrated commitment and focus.

Now all we needed was a ship.

Southern Quest

W hy we needed a ship: The southern polar summer begins in December and stretches to the end of February. This is expedition season, the only time when weather and sunlight provide conditions that are merely inhospitable, instead of downright lethal.

Antarctica plumps itself up in winter like an Arctic hare. It puts on another coat, another layer, not of fur but of ice. In the South Polar summer, Antarctica measures 5.2 million square miles. In winter, the ice pack doubles its size, to 10.4 million square miles.

The ice pack that surrounds the continent does not respect the season. It normally does not break up until midsummer, in late December or January. There is no way to get a ship close enough to land supplies until that time. But in order to have enough time to walk to the South Pole from the coast, an expedition has to start no later than November—the last month of the Antarctic spring.

When I tell the story of In the Footsteps of Scott, I invariably field the same kind of questions afterward. Why did you have to winter over? Why did you spend nine months in a hut waiting to go? Why did you need a ship to get you there? Accustomed to temperate-zone ease, many people have a hard time grasping the exigencies of polar travel.

Our schedule was being squeezed by the weather from both ends. We had no access to the continent by ship until January. We had a strict departure date for the expedition two months before, in late October or early November. In November the summer made it warm enough to walk. But the warmth did not break up the ice to allow ship navigation until at least two months later. If we waited for the ice to clear, it would be too late to begin our trek. The only possible way to make it work was to winter over. We had to spend nine months in Antarctica waiting for the right time to start. This is what Scott had to do, and this is what Footsteps of Scott was forced to do, too.

Why use a ship at all? Why didn't I just book a convenient airline flight? Land in November, in springtime weather, and start our bloody walk?

It sounded good, but it turned out to be impossible. There were no commercial flights to Antarctica, of course. The only ski-equipped air-planes were owned by governments. We had over a half ton of supplies for our trek to the pole, and we had ourselves, and there was simply no aircraft, commercial, charter, government, or otherwise, that was ready to take us.

You have to understand that Antarctica was a closed shop back then. Scientists only, please. "If I help one, I'd have to help them all," said Dr. Edward Todd, director of the American National Science Foundation's Office of Polar Programs in response to my request for support.

A private expedition, not sponsored by a government-endorsed scientific organization such as the BAS or the Scott Polar Research Institute, was just not done. The only presence in Antarctica was government-supported, not private, and governments were jealous guardians.

It was, like so many situations in life, an issue of control. A bureau-crat panics when faced with variables beyond his authority. Or perhaps the governments just didn't want anyone to see what was going on down there. Most of the research stations were surrounded by huge garbage dumps.

To be fair, the science bureaucrats who denied us cooperation might have seen In the Footsteps of Scott as a wedge that would open the door to Antarctica to all sorts of mischief. It is difficult to under-stand, nowadays, when boatloads of adventure tourists travel to the continent each year, just how isolated Antarctica remained in the mid-1980s. The great influx of seasonal visitors to the continent was, at that point, decades away. The expansion in polar bases, the boom that oc-curred during the cold war and before the implementation of the Antarctic Treaty System, had contracted in the course of the 1970s and early 1980s. Antarctica's population was falling, not rising.

More people had stood on the moon than had walked to the South Pole. As members of a private expedition, we were strangers attempting a journey to a strange land. No one would ferry us there.

One after another, we investigated options and saw them prove to be dead ends. We found an Australian company that chartered the huge ski-equipped Hercules airlifters designed for supplying polar stations, only to see the firm go belly-up. No one else flew Hercules charter flights anywhere in the Southern Hemisphere.

So we needed a boat. More properly, we needed a ship. A good-sized ship, too, to brave the harshest seas on earth. To give some idea of the conditions, the latitudes on the way from South America were nick-named the Roaring Forties, Furious Fifties, and Shrieking Sixties. The roughest latitudes, the seventies, were so rarely visited that they didn't even have a nickname.

We would be going as far south as you could take a ship. That was McMurdo Sound's attraction, that was why Scott chose it, and why the New Zealanders and Americans located their big bases there. From McMurdo it is only nine hundred miles to the South Pole, as opposed to two thousand miles from other areas on the coast.

McMurdo Sound is three thousand miles from New Zealand. Antarctica is the most isolated continent, surrounded by an unbroken band of extremely ugly ocean. Its closest neighbor, South America, is over six hundred miles away. No large landmasses interrupt the seas at those latitudes, and the westerly winds circle the globe with nothing to block them.

We were going to take a ship into those seas. We were taking a ship as far south as you could take one. We were going off the map.

We also needed a ship because we were wintering over. That meant shelter, that meant a hut, that meant coal and diesel oil for a generator and food and supplies for the long months of the Antarctic winter. In the end, we wound up taking twenty-four crew members and sixty-four tons of cargo, a load that ruled out not only any small sailing yacht but aircraft as well (the capacity of the C-130 Hercules, a.k.a. the Hercs, the most common airplanes to supply the government Antarctica bases, was 45,000 pounds).

There was quite simply no other way to do it. We had to leave Britain in a ship in the Northern Hemisphere's autumn, to arrive in the Southern Ocean in December. Then we had to wait for the ice pack to break up and allow our approach to the coast, drop off expedition members and supplies, hightail it back northward before the ice pack caught the ship in its icy grip. Spend January through October wintering over. Begin the walk to the South Pole on, say, October 25. The plan had the added appeal of closely following the way Captain Scott had done it in 1910. We would follow not only in his footsteps but in the wake of his ship the *Terra Nova*.

There was no way our expedition could afford a ship, of course, so the only thing to do was to dispatch Pete Malcolm to go out and buy one anyway. Off he went, riding his oil-spitting Norton in the dead of winter, freezing his backside off, touring the shipyards and harbors along the eastern coast of Britain. North from London, Lowestoft, Hull, and Newscastle. He found ships in abundance, idle and for sale. The British fishing industry was getting the stuffing kicked out of it by foreign competition. But most of the trawlers offered were ancient rust buckets. The few that were suitable were too expensive.

Pete would sputter up on the Norton, covered in ice, his arrival announced by a series of farting, banging backfires. He'd have to thaw out before he could talk. Then this oil-spattered, half-melted abominable snowman would announce that he was in the market for a ship . . . to go to Antarctica. No wonder few harbormasters or ship brokers took him seriously.

Farther north Pete the freezing motorcyclist rode, to Aberdeen and, finally, to a tiny port called Fraserburgh, in northern Scotland. There he found a fishing trawler originally named *Yesso,* recently refurbished and rechristened as *Cleanseas I.*

Andrew Salveson and Peter King at a shipping and transport company called Christian Salveson, which owned the *Cleanseas I,* took pity on Pete. The Salveson company had a long history. It was founded in the mid-nineteenth century but prospered on money from pre–World War II whale-hunting forays to Antarctica. So when Pete pronounced the name of the continent, it was as if he'd said a magic word.

I fielded a call at the Footsteps of Scott warehouse offices. Pete Malcolm's teeth-chattering voice came on the line. "I think I found us a ship," he said. He sounded cold but excited.

The price quote was fifty thousand pounds. Pete King would take a single pound as down payment to hold the ship. Peter Malcolm offered to pitch in twenty thousand pounds of his own money. I would put in everything I had. We begged our great friend Mark Fox-Andrews to make up the balance, which he most generously did.

Pete Malcolm had a navy man's enthusiasm for using the right vessel for the right job. The shipping part of the expedition equation was always his dream. He never sought to trek to the pole himself. Unlike so many of the hulks he had looked at, the *Cleanseas I* was ready to go. There were blankets on the berths and dinnerware in the galley, he told me.

I caught Pete's excitement and at the same time felt an odd sinking feeling in my gut. Footsteps of Scott was taking one giant leap toward becoming a reality. Be careful what you wish for. Could I really do it? It wasn't even our first night, and already I was having first-night jitters.

On the way up to Scotland to meet Pete in Fraserburgh, I calmed myself by running through possible names for our new flagship. My mother remembered that *Quest* was the name of Shackleton's last ship.

Perhaps if I had been more superstitious, I would have thought twice about her suggestion. Shackleton died aboard the *Quest,* in the middle of making a fairly senseless attempt to circumnavigate Antarctica.

Though sailing around the continent might have been meaningless, it had never been done before. But that wasn't the real reason why Shackleton wound up on the *Quest.* He was deeply in debt. He simply needed to mount an expedition, any expedition, to keep funds coming in.

Polar expeditions are so ruinously expensive that explorers usually can't stop at robbing Peter to pay Paul. They eventually must rob Paul, too.

I could have brooded on the fate of my predecessors, polar explorers who all died penniless. Instead, I wrote to Lord Shackleton, the explorer's son, to test out his feelings about our using the name of his father's last ship.

I thought *Southern Quest* a fine name for the vessel that would take In the Footsteps of Scott south.

Mile Eighteen

I was a man pushing a heavy cart up a hill. Then, almost without noticing, I reached the crest, and suddenly the momentum of the cart took over. I still pushed, but really I held on for dear life, trying to keep the pace while the cart rattled downhill, picking up speed. In the Footsteps of Scott took on a massive life of its own. I worked very hard just trying not to get run over.

But I was a madman. My commitment was total. I had no thought beyond the single idea of walking to the pole.

Marathon runners talk about "hitting the wall" at mile eighteen or so, a point where physical and mental exhaustion overwhelms many competitors. Successful marathoners must be able to find reserves of strength and determination to bust through the wall at mile eighteen and finish the race.

I was stuck in mile-eighteen mode. Overdrive, with a *Saturn V* rocket up my backside. For the seven months between the *Southern Quest* crashing into Irongate Wharf—I'm sorry, between the time the *Southern Quest* demonstrated its superb ice-breaking capabilities at Irongate Wharf—and the day we departed on the voyage to Antarctica, my legs, mind, and mouth churned as though they were runners determined to get past mile eighteen.

Money came in. Checks of twenty-five dollars and five thousand dollars. As Scott had done, I toured Britain, raising funds by giving lectures. My sister Lucinda helped immensely, since she ran a speakers bureau. My other sister, Rebecca, a lecturer on fine arts, helped me with public speaking. "No notes, Rob," she said. "Just passion."

I spoke at my old school, Sedbergh. I spoke at community centers and clubs. I spoke to whomever would have me. Will Fenton, cigar in one hand and phone in the other, worked tirelessly from a desk in the chilly warehouse office, getting the word out to newspapers, radio, and television outlets.

I recalled Scott, out on the lecture trail, raising money for the *Terra Nova* expedition and writing despairingly in his journal.

> Between £20 and £30 from Wolverhampton . . . £40 to-day . . . nothing from Wales . . . this place won't do, I'm wasting my time to some extent. . . . I don't think there is a great deal of money in the neighborhood . . . things have been so-so here . . . I spoke not well but the room was beastly and attendance small . . . another very poor day yesterday, nearly everyone out.

The most expensive lecture I ever gave, in a sense, was a free one to the kids at a local Docklands school near our warehouse. It was a grimy, sooty neighborhood, about to be obliterated to make room for the massive Canary Wharf redevelopment project. Though I spoke for free to the young, elementary-school kids, it wound up costing me millions of dollars, since after the talk I knew beyond a shadow of a doubt that I had to follow through with In the Footsteps of Scott.

My lecture was about Robert F. Scott. Presenting a talk, you can always tell when you are reaching someone. This or that audience member simply looks plugged in. At this talk, a slightly Dickensian boy with bulging eyes sat atop a desk at the side of the room, hanging on my every word.

He approached me afterward. "I really wish you luck," he said in a heavy East End accent. I half expected him to address me as "guv'nor." Then he pressed a fifty-pence piece into my hand.

The boy was poor. I knew it just by looking at him. He was slightly grubby, but his face shone with the fervor of the converted. The coin was an old silver piece, struck in honor of the Queen's twenty-fifth wedding anniversary. I still have it.

That fifty-pence piece was like a dagger to my heart. I thought about the boy with the bulging eyes often afterward. He came to stand in for the hundreds of schoolkids who had donated coins for In the Footsteps of Scott.

Why walk to the South Pole? I have had to answer that question constantly over the years. But the truth was, the why of it was still changing, still shifting.

The expedition began with the inspiration of the John Mills film and Scott's journals. It gained weight with the encouragement of my family. The effort became real with the backing of Sir Peter Scott and Lord Shackleton. Then people like Rebecca Ward and Will Fenton and Pete Malcolm gave me a new why. I found more and more that I was doing it for them. To honor their commitment. It wasn't just a dream anymore. It was my life. I was doing In the Footsteps of Scott because that's who I was, that was what I was doing, and so many good people had joined me on my quest.

But the damned fifty-pence coin drove the point home in a whole new way. I was stuck now, thoroughly nailed to my promise to go through with it. Okay, I might be a sentimental sucker. But I could never afterward pack up and skulk away from the expedition, because those bulging eyes would follow me. I couldn't let the coin boy down. All the schoolkids, all the coins, all the support, changed the nature of my why once again.

The most astonishing addition to In the Footsteps of Scott was a Docklands neighbor named Emma Drake. Everyone liked her. She was like the office's fairy godmother. She'd bustle in, bringing homemade elderberry wine, woolen socks or tins of goose grease, which, she said, would protect our faces against the cold.

Emma was psychic. She predicted we would reach the South Pole on January 11, which we did. I never really took her clairvoyance seriously, but once, when we were down to our last penny and convinced a sponsor's promised infusion of funds would never arrive, Emma predicted, "You will get the check." That day a sponsor's check arrived in the mail from a most welcome supporter named Jack Hayward. I learned later that Major Cordingley had had a quiet word with him to encourage his support.

Roger Mear also joined the effort at Metropolitan Wharf. At first the expedition was going to comprise just Roger and me, a two-man team walking to the pole. Roger refined our vision. My original idea had been to duplicate Scott's expedition. As Captain Scott had taken

Siberian ponies and sled dogs, we would take ponies and dogs. We would depot supplies just as he did. Roger zeroed in on what would be possible and also what would interest him as a professional mountaineer. He obsessed over the idea of being unassisted. That meant no dogs, no ponies, no depots, no supply drops. Just two men hauling everything they would need, food and fuel for three months, pulling sleds for the nine hundred miles to the pole. He wanted us to respect the isolation and commitment of the real polar explorers. "As far as I can tell, it would be the longest unassisted march in history," Roger said.

Mounting an expedition, you needed to be able to say something like that. Stake a claim that set your expedition apart from all the other expeditions out there, and all the others that had come before. "The world's first . . ." "the longest . . ." "the highest . . ." It was part of the game. A rubric to march behind, like a flag. We needed something for Will Fenton to say and for journalists to write down.

The longest unassisted march in history. It was true. Plenty of people, from Napoléon to Mao Tse-tung, had made longer marches. But no one had done anything close to what we were doing, not unassisted. Our forerunners on long treks could always live off the land. In our case, there would be no bounty to plunder, no fat of the land in the frozen wastes of the barrier. After we left the Antarctic coastline, there would be no seals, no penguins, not even any birds once we were deep within the continent. No food for which to forage. No radio communication.

Perhaps T. E. Lawrence's famous crossing of the Sinai Peninsula from Aqaba to the Suez Canal is, in miniature, a more apt comparison. The Sinai was a provisionless desert, which Lawrence crossed in forty-nine hours, without sleep. Our trek across Antarctica would last three months. Roger liked the solitary, isolated challenge of it, man against the elements.

I focused on the parallels to Scott. With an assembled caravan of sixteen men, ten ponies, two teams of dogs, and two motorized sledges, he departed on his pole march on November 1, 1911. The expedition broke into detachments, each running at a different rate. "It reminded me of a regatta or a somewhat disorganized fleet with ships of unequal speed," Scott wrote.

The innovative motor sledges were ahead of their time and broke down almost immediately; the dogs and ponies struggled and died; the five men who went on to the pole perished also.

Not infrequently, I questioned the wisdom of following in the footsteps of an expedition that eventually turned into a death march.

"Show me a hero," said F. Scott Fitzgerald, "and I'll write you a tragedy."

F. Scott was certainly correct about Robert F. Scott. But was tragedy inevitable? I think I preferred my own personal heroism straight up, without the chaser of martyrdom.

The press liked both the historical parallels and the unassisted pursuit of our goal. In the Footsteps of Scott reached a critical media mass over the summer of 1984, to the point where a large segment of the public was aware of it. Reporters loved to call us "mad." We were the "two mad Englishmen who are going to walk to the South Pole."

Sitting at their warm hearths in front of a cup of hot tea, readers and viewers relished the prospect of a horrible march across an icy wilderness, as long as somebody else was accomplishing it. "Horror journeys," said war reporter Martha Gellhorn, "are the only ones we really like to hear about." (Or Paul Theroux: "The worst trips make the best reading.")

The converse is also true. Whenever I'm on expedition, it is always very satisfying to dwell on the image of the people back home. Somehow, the idea of a fireplace always figures in, or more particularly, one that I always seem to summon up with the phrase "a warm hearth." Comfort and laziness. Heat and inertia. In gruesome conditions, the warm hearth represents a consummation devoutly to be wished for.

Our increased public profile paid off in attracting patrons. Sir Vivian Fuchs, head of the first expedition to cross the Antarctic Continent, signed on, as did such luminaries as Wally Herbert, the first man to cross the Arctic, and Lord Hunt, expedition leader for Edmund Hillary's successful first ascent of Mount Everest.

The toughest nut to crack turned out to be Lord Edward Shackleton, son of Scott's sometime subordinate and eventual rival, Ernest Shackleton. On first contact, he turned me down flat. I kept pestering him with letters and phone calls. I believed that the Shackleton name, paired with Peter Scott's atop the expedition's patron list, would show that old wounds were now healed.

Lord Shackleton finally came around, and became one of our most dogged supporters. Citing the experience of his father, he convinced us that a two-man expedition was quite mad. He insisted that three would be a lot safer. Roger and I decided that Gareth Wood would be our third.

Just as essential, we attracted a professional to organize and streamline our fund-raising: Richard Down, owner of a company called Interaction, which promoted sports events and adventure enterprises. Without Richard, I think I would still be at that dock on the Thames today, waiting to depart for Antarctica. He kept the funds coming in the door, but as fast as they came in, they went out again. The *Southern Quest* proved the truism known to every sailor, about a boat being a hole down which you throw money.

Lord Shackleton joined me for a meeting with John Raisman, chairman and CEO of Shell U.K. Shell was Barclays Bank all over again. I had called the company repeatedly, suggesting that support of our expedition would honor the historical association of Shell supporting Captain Scott's.

But we did not support Captain Scott's expedition, Shell's publicity people informed me.

You sponsored Captain Scott, I said.

No, we didn't, Shell said.

So that's where we stood. We had a ship but no fuel to run it. I recalled that somewhere, at some point during my research into Scott's mission, I had seen a photo of him sitting on a wooden box marked with a company logo and the words SHELL SPIRIT. But when I looked back over my notes, I could not find a reference to it.

"I have to find that bloody photo!" I bellowed, knowing that a Shell sponsorship might hang in the balance.

Lord Shackleton knew Raisman and set up the meeting at the Shell Mex House on the Thames.

Lord S. spoke first, astonishing me by launching into what sounded like an attack on my experience, my motives, and my background. "You know, when I first met Robert I thought his expedition was totally harebrained," Shackleton said, puffing on a big cigar.

I sat there, red-faced, as he tore me to pieces. I was arrogant, he said. Inexperienced. A foolish amateur. Puff, puff. I almost got up and left the room. What had he lured me here for? Was it just a trap to put me so thoroughly in my place that I would never venture into public again?

But Lord S. skillfully changed his tone. "Gradually," he said, "I became quite impressed by Swan's determination and acumen."

By starting out with a full-scale assault, he had forestalled any objections that the chairman of Shell might have to me, a young upstart coming to him hat in hand.

"I think Robert has something to show you," Shackleton said, finishing his introduction.

I brought out the photo of Scott sitting on a box of Shell petroleum. Learning how to do research properly at the University of Durham paid off. Searching through file after file at the British Museum, I had finally found the photo and had it copied.

I left the Shell Mex office with a commitment from the company to provide for all the fuel requirements of In the Footsteps of Scott, including more than six hundred tons of diesel—in effect, a $200,000 contribution to the cause.

Summer 1984 blurred into early fall. The *Southern Quest* received hull strengthening to render her suitable for Antarctic waters. Dozens of volunteers, paid very occasionally, and then only with room and board in the drafty warehouse, helped out on the ship. She was fitted with ice-deflecting plates on her rudders. She was scrubbed and stocked and outfitted. Our expedition supplies burst our warehouse headquarters at the seams, and we created an annex down the Thames, Shed 14 at West India Dock—the same wharf Scott had used on his expedition.

Roger focused on the polar journey itself, navigation, logistics, equipment. Gareth concerned himself with the hut we would erect at Cape Evans on the shores of McMurdo Sound, as well as the supplies for our entire winter over. As a test, he erected the prefab hut inside Shed 14.

We also aimed our efforts on the *Southern Quest* toward getting a Lloyd's of London Ice Class III approval to qualify for insurance. Finally, though, at the very last minute, the underwriter balked.

I had a choice. I could cancel or delay the expedition. Looking around at the volunteers in the shipyard, somehow signed on to my dream with no hope of compensation or even chance of going to the pole, I simply refused to face defeat. Pulling out was not an option.

So I underwrote the ship myself. With the heroic support of Don Pratt of Barclays in the Strand, I mortgaged my house and paltry belongings, then signed a personal guarantee for the rest. To the tune of $1.2 million.

"Don't worry," I assured the obviously very worried bank manager. "I'll sell the boat when we return, and Barclays will get its money back." It was a promise that would come back to haunt me.

The *Southern Quest* wouldn't be the first vessel floating away from port on a tide of red ink. To avoid creditors, Roald Amundsen sailed in secret out of Christiana (Oslo) on his successful Antarctica mission to conquer the South Pole. Scott's own competing trip aboard the *Terra Nova* was woefully underfunded. When the *Southern Quest* finally departed Britain, I'd be joining a long tradition of polar expeditions in debt up to the eyeballs.

As it worked out, I wasn't even on the ship as it left port. All we needed now was twenty tons of top-grade coal.

Cape Town (II)

Captain Scott himself was rarely on board when his expedition ships departed from home port. There were too many last-minute details to attend to, too many funding sources to cajole, too many speeches to make attempting to shore up support. When the *Terra Nova* sailed from Cardiff, Wales, on June 15, 1910, Scott stayed behind, taking a faster packet boat a month later to catch up with his ship in Cape Town, South Africa.

Historical parallels aside, I would have been overjoyed to be aboard the *Southern Quest* upon her departure from London. Just as Scott had been, I was becalmed by details, mostly of the money kind. When you mortgage yourself to the tune of $1.2 million, people make you sign a whole raft of papers. An old line from Bob Dylan kept running through my head: "They asked me for collateral and I pulled down my pants."

I did manage to make a celebratory dinner in Cardiff, the *Southern Quest*'s last port of call in Britain, just as it had been for Scott's *Terra Nova*.

The Captain Scott Society of Cardiff hosted a banquet for us in the same room where Scott's farewell banquet had been held, with the same menu, even the same music. This was the banquet, infamous in Scott lore, where Taff Evans, his petty officer, got himself so drunk he fell into the Cardiff harbor afterward.

I stayed upright (just), but the constant parallels with Scott were beginning to make me nervous. I realized I had set it up that way myself, that I had mounted an expedition called In the Footsteps of Scott, but with the twenty-five-course dinner and the harp music the evening began to feel a little creepy, as though we were the Ghost of Robert F. Scott expedition. The only suitable thing for me to do, obviously, would be to die on the ice as Captain Scott had done.

We were in Cardiff partly because we needed its good Welsh coal.

Wales produces some of the cleanest-burning anthracite in the world. Coal would be the safest method by far to heat our hut in winter. But the miners of Wales were on strike. No coal moved in or out of the port at Cardiff. I met with some top officials from the Welsh miners union. I am sure they expected me to be a snobby English git, and I tried extra hard not to be. "I completely sympathize with your cause," I said. "But we've come here expecting to take on coal for the expedition."

"We're on strike, young man," the topmost of the top mining-union officials in town told me, as if closing off debate.

"Isn't there something we can do?" I begged.

It didn't look as though there was. But that midnight, while we bunked on the *Southern Quest,* a truck pulled up on the wharf beside the ship. "We've got your coal," the driver told us. "But this never happened, you hear me?" He offloaded twenty tons of high-grade Welsh coal. We never told anyone, and we never got billed for it. Thanks, union dudes!

The *Southern Quest* cast off from Cardiff on November 10, 1984. Peter Malcolm made the voyage out. I stayed behind to be pecked to death by fund-raising ducks, and to dot all the *i*'s and cross all the *t*'s on loan documents. In early December I flew to Cape Town to meet the *Southern Quest.*

I got there before the ship. There was no satellite tracking in those days, no cell phone service, so I had no precise idea when the *Southern Quest* would arrive. I received a radio message that she was a day from port. Restless, wanting to keep in shape, I climbed Table Mountain, a promontory three thousand feet above Cape Town.

Reaching the top, I stopped, scanning the horizon. Nothing, no ships of the right size and shape. Spread out below me were my old haunts from ten years previous, when I drove a cab to earn money for a very long bicycle ride: Marine Drive, Table Bay Boulevard, Koeberg Road.

The Orwellian year of 1984 was just drawing to a close. As I waited on Table Mountain, it felt like the first time I had stopped running, churning, pushing, since I'd started organizing In the Footsteps of Scott a half decade before. I had been stuck in mile-eighteen beat-the-wall marathon mode for so long, I could not get unstuck.

The experience of those five years had served me well. Commitment is contagious. I had fixed on a single idea and allowed no other ideas to distract from it. I didn't start a family. I didn't launch a career. If I'd had anything besides the expedition in my life, I probably would not have been standing where I was, waiting for my ship to come in.

The deadline of the previous few months had been murderous. If the *Southern Quest* had been delayed in the United Kingdom any later than it was, we would have had to put off the expedition for a year and probably would have lost it all. So that was a second takeaway: Don't let a deadline kill your progress. In the face of a time crunch, never get overwhelmed, never give up. Power forward as though you are going to make it, and you will. Deadline anxiety kills more projects than it saves.

Another imperative of leadership I had wrung out of the chaotic days of preparation: Recruit good people, then get out of the way and let them do their jobs. This is essentially a question of control. Many people are queasy about giving up command and decision-making power. But if I had tried to look over the shoulder and second-guess everyone who worked to get In the Footsteps of Scott on its way, the result would have been killing. Literally. My killing, at their hands.

Then there was the Lesson of the Crash at Irongate Wharf. Bad publicity, it turns out, is sometimes good publicity. The embarrassing gaffe when we bulldozed the *Southern Quest* into the crowd of supporters and media turned out to be a blessing. I don't suggest that anyone go out of his way to commit such blunders, but I would point out that there's no way to tell at the time how apparent disasters will play out.

But my biggest takeaway was this: If you say you are going to do something, do it. No excuses, no dodges. That's probably rule number one of being a leader. Organizing In the Footsteps of Scott tested me on this conviction every step of the way, in a hundred different promises that I made to supporters, funders, and the team every day. But the overall test was that I had said I was going to organize an expedition to walk to the South Pole. I said it, and I had to follow through.

I realized that my why had changed once again. Why follow Scott? Because in quite a small way, I was becoming something. I was becoming the bloke who was walking to the South Pole. My dream reached critical mass. All around me, I saw the best minds of my generation, my friends, cousins, and schoolmates, joining the corporate world, getting married, having children.

Ordinary life was passing me by. I was somebody different, not run-of-the-mill, for doing what I was doing. Maybe only in my own eyes, but that was enough. I wasn't *trying* anymore. I was doing. Calling myself a polar traveler when I was younger always went over well with women I was trying to impress at parties. It pumped up my ego. I had talked myself right into the role.

Bit by excruciating bit, I had begun to understand what building a team and being a leader might entail. Because that was the true expedition I had embarked upon. Not outward to the pole but inward, into myself. What was I doing with my life? What are any of us doing here? What are we doing with our time? I thought of the line from an old Inuit poem: "What is this life of mine?"

Scott, Shackleton, and Amundsen had all gone through the same process. I clung to them as my models. Three long-dead heroes joined my grandparents, my parents, my siblings, my history teacher at school, an actor in a film, the Sedbergh ten-mile run, a wild-haired motorcyclist, a donation of three thousand cans of beer, women at parties, and a goggle-eyed schoolboy with a coin in his hand.

They had all led me here to Table Mountain, where I stood waiting for my ship to come in.

I had been staring westward into the flat expanse of the Atlantic Ocean, over Cape Town Harbor, the Cape Peninsula, and Table Bay. Far to the west, I saw an orange-red dot. My field glasses hung around my neck. I brought them up and, after a little fiddling with the focus, saw the *Southern Quest* throwing up white wavelets of foam as she powered her way into port.

I had known it all along in the abstract, but as I looked out from Table Mountain it became real. For the first time in my young life, I was at a place where I couldn't talk my way out. As the red dot on the ocean grew closer, a few details resolved themselves: the trawler's white superstructure, its radar antenna, Pete Malcolm's hair.

Well, not really, not that last one. But as the *Southern Quest* neared port I did catch a glimpse of the beautiful Rebecca on board, now in her role as expedition photographer. I realized I had so many people depending on me.

No going back, I thought to myself. *I am going to have to walk to the South Pole.*

My language was a little more colorful than that, because the reality hit me with a force that was not, truth be told, 100 percent pleasant.

The Monty Python March

The *Southern Quest* lodged in dry dock in Cape Town. A shipbuilding and refitting company called Safmarine had taken the expedition on as a pro bono project. Massive supertankers also in the dock dwarfed our little ship. It resembled a child's pony in a herd of Clydesdales.

December 1984 was a dark time for South Africa. It remained an international outcast for the government's senseless adherence to apartheid. I had been counseled by many people to have nothing to do with the country. *Do not go there. You will only be granting legitimacy to an evil regime.*

I understood the argument, but I believed active engagement was always the better way. In future years, I would journey to China when there was similar pressure against that country for human-rights abuses. I would work with coal companies on energy initiatives, much to the disgust of my purist friends. Engagement, even if it involves compromise, always strikes me as the most effective way to foster progress.

Besides, Scott had stopped in South Africa on his way to New Zealand.

At the Safmarine dry dock, a hundred workers swarmed over the *Southern Quest*. In record time, they blasted through every repair the ship needed. I was stunned. I felt like a bystander. Where did I fit in? I caught some strange looks from the longshoremen, sailors, and harbor girls I had encountered driving a taxi ten years before. I felt a little bewildered by the decade's worth of changes.

"You must change your life," Rilke wrote. In the throb and heat of the *Southern Quest*'s engine room, I decided that was just what I had to do. For the previous five years I had functioned first and foremost as a salesman. I sold my idea, my dream, my expedition, to anyone who would listen. Along the way I realized that one person I was selling it to was myself. I had to be sure, thoroughly and irrevocably sure, that this was what I wanted to do.

Now that I was aboard ship, none of that meant anything anymore. Who was I selling to now? Still to myself? No, the moment on Table Mountain had represented the final sale to that particular customer. The other crew members? They were sold on the expedition already, or they would not have been aboard. I had to change my life from being a person selling the idea of walking to the South Pole to being a person who could indeed walk there. This required a fundamental shift in head space.

The other expedition members had enough experience that they could at least imagine themselves on the march. Roger Mear had dangled inside a mountaineer's sleeping bag on the sheer rock face of the Eiger, letting winter winds howl him to sleep. I had done nothing remotely in that league. In preparation for the trip, and to give myself at least a hint of credibility, I had skied in Iceland, man-hauling my supplies. But that was it. I was going to be in a state of complete panic on the expedition if I somehow could not force a change in my thinking.

As a first step, I had decided to work in the ship's engine room until we got to our destination. That meant the duration of the voyage from New Zealand, where I finally joined the *Southern Quest* at our jumping-off point for Antarctica. So it was belowdecks, from Lyttelton south to the continent itself. The salesman-figurehead for the expedition might more properly be topside, posing for pictures. Being down in the heat and noise was a way of saying, to myself and others, *I am not that guy anymore.*

The oil, grease, smoke, and dirt of a shipboard engine room mucks everyone who works there. I would emerge exhausted and filthy and flop down into my berth to catch a few hours of sleep. I know it changed the crew's opinion of me. I can just imagine the resentment I could have caused if I had retained my picture-taking role.

But more important, working belowdecks changed my idea of myself. I didn't have a clue about navigation. There was no real reason for me to hang about the bridge. Captain Graham Phippen, John Tolson, and Peter Malcolm, the sailors on the expedition, were the ones who were going to get us there safely.

Recruit good people, then get out of the way and let them do their jobs. The engine room was as out of the way as I could get.

Before leaving Lyttelton, New Zealand, I had made an incredible connection to the original Scott expedition. A ninety-six-year-old gentleman named Bill Burton, the last surviving member of Scott's crew, met us quayside. As a young man, Burton had shoveled coal on the *Terra Nova*. Shaking the hand of a man who had shaken Scott's hand felt like a passing of the baton.

Everywhere in New Zealand, I felt echoes of Scott. A heroic sculpture by Kathleen Scott of her husband stands in a Christchurch park. I felt vindicated, recalling the last time I had been in the city, sleeping under a tree before begging Bob Thompson to allow me to join an expedition with the New Zealand Polar Division. I said I was going to do something, and here I was back again, well on my way to doing it.

In the Footsteps of Scott was getting a cold shoulder not only from the New Zealand polar establishment but from the rest of the international community as well. The United Kingdom's prestigious Royal Geographical Society had given us its seal of approval only hesitatingly, under the prodding of our patrons, Peter Scott and Lord Shackleton. Officials from the United States, a major presence in Antarctica, went beyond reluctance to downright hostility.

America maintained bases at both ends of our trek, on Ross Island, where Scott had wintered over in a hut on the shore of McMurdo Sound, and at the South Pole. We had not consciously set out to rub shoulders with the U.S. scientists. Far from it. We would have preferred to avoid all human population. But history had directed us to McMurdo, and the Americans happened to have built their bases precisely where Scott walked.

The remarkable U.S. base at the pole marked our finish line. Maintained by seasonal (October through February) thrice-daily flights of ski-equipped Hercs, the Amundsen-Scott South Pole Station represents a strategic statement by the United States. Its placement at the center of the polar pie means that America could conceivably claim sections of all the slices of the continent.

No one owns Antarctica. In that respect, it is unique in the world. The Antarctic Treaty System (ATS), first signed in 1959, with environmental protocols added in 1991, reserves the continent for peaceful scientific research, banning military and commercial exploitation. The ATS was designed to prevent jockeying for position and territory on the part of world nations.

The jockeying began soon after the Heroic Age of Antarctic Exploration closed and reached its height in the cold war. The United Kingdom, New Zealand, Australia, Argentina, Norway, Chile, and France eventually asserted territorial claims, and Russia and the United States erected bases.

During World War II, Germany was eager to get into the land grab, even though it lacked the resources actually to construct an Antarctica station. The Nazis dispatched a plane to fly over the continent and drop hundreds of small metal swastikas as a way of marking the claim for German sovereignty.

Admiral George Dufek of the U.S. Navy won the "second race to the pole" in the year of my birth, when in October 1956 he and six others set up the base that would eventually become the Amundsen-Scott South Pole Station. The Soviets responded by setting up Vostok Station at the "pole of inaccessibility," latitude 78°28' S, longitude 106°48' E, the most remote point from the Antarctic coastline.

Since no one owns Antarctica, no one could officially prevent In the Footsteps of Scott from doing what it prepared to do. But de jure and de facto are two different concepts, and the U.S. officials did everything in their power to discourage us. Dr. Edward Todd, of the U.S. National Science Foundation Office of Polar Programs, stated directly the impossibility of cooperation. For our return from the pole after our expedition, the United States denied us space in the cargo hold of its big Hercules airlifters, even though the planes normally flew back empty on their supply runs to the Amundsen-Scott Station.

Our march was to be "unassisted" with a vengeance. After I got over my naïveté regarding international politics, I actually came to relish our independence. Never underestimate the spirit of "screw you" as a power of motivation.

I fully recognized that such luminaries as Dr. Todd and Bob Thompson of the New Zealand Polar Division had their precise agendas to follow. Antarctica was a "natural reserve for science," and I respected that. But no one had the authority to tell us that we could not mount an expedition to honor Robert F. Scott, Ernest Shackleton, and Roald Amundsen. It was not our fault that the path of our trek lay directly athwart two massive American bases.

I knew where the governmental polar divisions were coming from. But I felt they needed to know what we were all about, too. We were taking along no radios. We were marching without assistance. That meant we were fully intending not to require theirs. I felt the quid pro quo involved was dead even.

On February 2, 1985—they were celebrating Groundhog Day in America—the *Southern Quest* slid out of its New Zealand harbor, towed by a restored steam tug called *Lyttelton*. The same tug had towed Captain Scott's *Terra Nova* out of the harbor almost seventy-five years earlier, in November 1910.

A brass band from Skellerup Industries (one of our sponsors and a leading purveyor of gum boots) played a Sousa tune, the one John Cleese and his out-of-their-minds troupe of comic friends have taken as their theme. The tooting notes of "The Monty Python March" sent us on our way.

We were on our own.

Jack Hayward Base

McMurdo Sound extends some thirty-five miles inland, a blunt finger poking south off the larger Ross Sea, with the mountains of Victoria Land and the Scott Coast bordering it. To the east rise the black cinder spurs of Ross Island, topping out with Mount Terror and its more impressive twin, an active, smoke-plumed volcano, Mount Erebus.

February. The dog days of the uncertain Antarctic summer, and the only period of the year when McMurdo Sound was navigable. The *Southern Quest* arrived without fanfare in the most famous waters of the Southern Ocean.

Farther south, at the base of the sound, the big American base called McMurdo Station passed for the bustling New York City of the continent. With its nearby sister facility, New Zealand's Scott Base, the area was by far the most populated hub in Antarctica. Between us and it, the Erebus Glacier Tongue reached across the sound.

Here, twenty miles from McMurdo, all was silent. Not a single human footprint showed on the thin, stony beach as we slid past, just crabeater-seal tracks. Straining with anticipation, Pete Malcolm and I had climbed into what passed for a crow's nest on the *Southern Quest*—a Vaux beer barrel affixed to a masthead.

Beneath Erebus's sulphurous plume of smoke, a beautiful gold-rose band of sky showed itself on the horizon. On the trip down from New Zealand, the ship had been constantly shadowed by birds, as though on the lonely Southern Ocean any company was welcome, even that of a rust-colored trawler. Storm petrels mostly, and plunge-diving gannets, but as we got farther out to sea, the albatross. Coleridge's omen bird.

We spotted many different species: wandering, royal, sooty albatrosses. On my few forays onto deck from self-exile in the ship's bowels, I found the big royals mesmerizing to watch. They rode the air currents above the waves, dipping to feed, gliding and soaring but flapping their wings only rarely.

Some birds fly to live, other birds live to fly. *That's how to do it,* I thought, watching the albatross. *Use the forces in the environment around you. Don't expend a shred of extra energy. Glide. Don't force it.* The best way to get by in the harsh polar environment of Antarctica (and maybe everywhere else, too).

The albatrosses dropped away as we entered deeper into McMurdo Sound. Now it was mostly skuas, quarrelsome, squawking birds, the pigeons of the polar regions. Great skuas, nicknamed bonxies, haunt the ports and beaches of northern Britain. Some smaller species, called jaegers, are Holarctic, and I had encountered them in Iceland. These skuas in the south were huge, ornery, and wont to dive-bomb. A small flock of brown skuas had already checked out Pete and me in our crow's-nest barrel. *What are you doing up here?* they shrieked. *Hell if I know,* I replied.

Then, appearing suddenly off the port bow, the real answer to that question appeared. Scott's hut. A small wooden building of weather-beaten gray, set a short distance from the water's edge. The shoreline jutted weakly into the sound just there.

Cape Evans. Where Scott wintered over before he made his fateful trek to the South Pole.

I was prepared for it. I knew it would be there and had visited it many times before in my mind, but actually seeing it for the first time overwhelmed me. Pride that we had made it mixed with relief and a shade of terror over the desolation of the place.

"There it is," Pete said quietly. The skuas screamed like madness overhead.

Frozen in time. A three-word cliché that on the shores of Cape Evans was quite literally true. There are few decay microbes in the Antarctic ecosystem. Normal processes of decomposition simply do not occur. Scott's hut was as he'd left it. In the crow's nest of the *Southern Quest,* it was 1985. On the beach, five hundred yards away, it was still 1911.

History implies change. I held a degree in ancient history and had arrived at a place without history. Antarctica's human chronology began in 1820, a mere eye blink in the stretch of time. No Antarctic languages developed, no folk songs were created, no mythologies, no traditional cultures at all. A tabula rasa. Antarctica was sterile in more senses than one. Whatever was there was what we had brought with us.

Which turned out to be quite a bit, as we drove our mooring ropes into the ice edge three hundred yards off the black-lava beach at Cape Evans and off-loaded the *Southern Quest*. We had ton upon ton of supplies, equipment, and provisions. Six hundred sacks of strikebreaker Welsh coal for heat during the dark frozen months of winter. Barrels of Shell-donated diesel for the electrical generator. Crampons and snow gaiters and sleds for the march to the pole. Lumber and prefabricated parts for our hut, which we would erect two hundred yards down the shoreline from Scott's ghostly spook house.

Giles Kershaw, the great polar pilot, had given us a piece of advice that demonstrated extraordinary foresight. Bring ashore ten barrels of jet-1A aviation fuel, he said. "You can use it for your stove," he told us. "It's just the same as kerosene."

We have stove fuel, I told Giles. Why do we need aviation fuel?

"You never know," Giles said. We didn't realize it then, but those ten barrels of fuel would turn out to make all the difference in the world.

We landed human cargo, too—the equipment and personal effects of the five of us who would winter over in the hut. Myself, Roger Mear, Gareth Wood, Dr. Mike Stroud, and Captain John Tolson. Gareth had been recruited for the expedition by Roger, and Mike and John I had met along with Roger at the Rothera base.

As it turned out, those five fell into natural roles. Roger would get us to the pole. Mike would make sure we didn't die. Gareth would organize. John would make a film and take photos. It was now time for me to keep my mouth shut and follow the leadership of the others.

But we made a crucial mistake in organizing the trip. Out of the five who wintered over, only three would walk to the pole. Roger and I had been in from the start. We had assumed that Gareth would be the third. But personal difficulties were arising between Roger and Gareth. Might Mike be the better choice?

Which would it be? Mike, as a physician? If something happened during the trek, wouldn't it be vital to have a doctor along? Gareth had extensive mountaineering experience. The fact that Roger had asked him along on the expedition spoke for itself.

Transferring ourselves and our mounds of equipment from ship to shore made for strenuous, brutish work, harder than any I had done as a base-camp general assistant for the BAS. The *Southern Quest*'s windlass was broken, rendering the task even more arduous.

Our chippie, or ship's carpenter, Mike Seeney, constructed a slapdash pontoon boat to aid the off-loading, lashing together barrels and boards and christening the rickety result the *Spirit of Incompetence*.

The pressure of changeable weather loomed over us. McMurdo Sound iced over quickly, at notoriously unpredictable times of the year. Winds blew pack ice into the sound off the Ross Sea, blocking it to navigation. Humping sack after sack of coal until I again took on my soot-grimed Berber disguise, I obsessed about getting the ship unloaded. She had to be on her way north before ice locked in the whole sound for the winter.

I had a last vital duty on board: saying farewell to Rebecca Ward.

She had been a true friend. Antarctica wasn't her thing, it was mine. But she had soldiered alongside me. When it came time for me to leave, Rebecca gifted me with a small stuffed bear that she had sewn, made of gingham and packed with scented lavender. "I want you to take him—" she said.

"Teddy," I suggested as a name, always the imaginative bloke.

"Yes," she said, laughing. "I want you to take Teddy to the South Pole. He's for all the schoolchildren who supported you."

I thought of the boy with the bulging eyes. She didn't need to tell me. "They're the ones," I said.

"Yes." Rebecca nodded. "They're the ones we're doing it for."

Our tears froze on our faces.

A week after we first approached Cape Evans, on February 17, the *Southern Quest* weighed anchor (a process made particularly backbreaking by the lack of a windlass, and which took five hours). Over the course of those seven days, we off-loaded sixty-four tons of supplies. The beach, which the winds kept scoured clean of snow and ice, appeared a chaos of boxes, barrels, and tents.

Mear, Wood, Stroud, Tolson, Swan. The five of us who would winter over at Cape Evans watched the *Southern Quest* disappear in North Bay of McMurdo Sound. Across the water, sun lit up the islands of the bay and Victoria Land's mountains, ridges, and glaciers. At the other end of our beach stood Scott's hut. Out of sight some twenty miles to the south lay McMurdo's big American and New Zealand bases.

Jack Hayward Base we decided to name our little corner of a very big continent. After Jack Arnold "Union Jack" Hayward, Bahaman real estate mogul, owner of the Wolverhampton Wanderers football club. Our benefactor Hayward had more than once rescued In the Footsteps of Scott with crucial last-minute donations when all looked lost.

We would spend the next nine months together at Jack Hayward Base, six of those months in the darkness of the Antarctic winter, confined for much of the time in a hut that measured 16 × 24 feet. With the interior filled with equipment and blocked out by a pantry, bathroom, and darkroom, our effective space would be more like 16 × 16.

It was probably too soon for us to start hating one another.

Wintering Over

A little over two months after we landed at Cape Evans, on April 23, the sun left the sky. Amundsen used to call the sun "His Grace," and when it leaves your life, you understand the metaphor.

The sun had been hovering near setting for weeks. It played odd tricks. As the polar darkness grew to engulf all daylight, the sun ceased to be a disk in the sky and transformed itself into an orange-gold column of light, a rectangular bar embedded in the horizon line. The stationary solar display recalled the impersonal black monoliths in *2001: A Space Odyssey*. Only this monolith was on fire.

By the time the Antarctic winter night began, we had feathered our well-insulated, coal-heated nest fairly well. The temperature in the hut cheated on the warm side and felt warmer because of the outside conditions. To cite just one sample reading, the temperature measured sixty-five degrees F inside the hut, when outside it had fallen to −23, with fifteen knots of wind. There were many days when I never ventured farther from the hut than the meteorological-instrument screen, a few yards outside.

We each made our own personal cubby in the space beneath the roof. I spent a lot of time on the typewriter, a manual model in those days before laptops. Even today I have a heavy hand at the keyboard, trained by a cold, sluggish machine, on which I pounded out thank-you letters to each of our hundreds of sponsors, patrons, and supporters.

I would mail the letters during one of our rare forays to New Zealand's Scott Base, twenty miles to the south. The Kiwis ran a working international post office, it being good for their sovereignty claims to do so. We could manage to get at least a few letters out to family, friends, and supporters. When was the last time anyone got a letter postmarked Antarctica?

(Later in my life, in my visits to the same sponsors—probably in search of another round fund-raising money for some other outlandish expedition—I would occasionally stumble upon one of my letters hung on the wall of a hallway somewhere, next to a supply closet or on the way to the fifth-floor lavatory. My thank-you note, with its Scott Base P.O. stamp. "Oh, yeah, that's you?" the young employee would say. "I always wondered what that was. Never read it." How cruel history is.)

I wrote to John Mills, the actor. I told him about watching *Scott of the Antarctic* as an eleven-year-old. "This is all your bloody fault," I wrote, only half kidding. "I'm sitting in a hut a few hundred yards from Scott's hut. I haven't seen a woman in five months and won't for another six."

I was stunned when, a couple of months later, Mills wrote me back via Scott Base. He had sent a glamour-shot publicity photo from the film, his face with snow goggles on, totally covered in ice. "Dear Robert," he wrote, "if you don't look like this, you are probably going the wrong way."

The constant heavy clacking at the typewriter got on everyone's nerves. *Anything* any of us did got on the nerves of the others. I could have sat silently in my den, and that would have been equally annoying. In the Footsteps of Scott was an extreme endeavor, so it naturally had attracted extreme personalities. Our nuclear family exploded.

In the central space of our hut, a rounded sheet of marine plywood (fashioned by John Tolson) hosted most of our meals. Around this table the group's psychological dynamics played out. There was no place to hide. In the normal course of the day back at home, the elements of my character could be masked or cushioned. But in the claustrophobic space of the hut they became sharp edges against which the others rubbed.

In many ways, I found Roger Mear to be the polar opposite of me, so to speak. No matter what was said, no matter what action proposed, Roger had a one-word response: "No." I was a yes-man, finding it easier to say yes, even if I didn't entirely mean it. Easier than to put up with the hassle of saying no. Roger had the character to say no. But gloomy moods hung around his person like the smoke around the peak of Erebus, the volcano that we could all see from the north window of the hut. Roger was our own personal Erebus.

Gareth Wood drove me crazy and was equally driven crazy himself by the gap between his mania for cleanliness and organization and my general slovenliness. His diary revealed one episode when he tracked the progress of a dish towel, from the point when he neatly hung it, to when others used it and balled it up, to when he again neatly hung it, and someone else balled it up again, and so on. We were a group of smelly, farting, belching human males in cramped quarters. Any sense of order was relative.

To liven things up, we hosted a sadist in the bunch. Every two weeks Dr. Mike Stroud hooked us up to an electrocardiograph and jammed breathing masks on our faces, ostensibly to measure physiological responses to cold, but really to delight in our discomfort. The only redeeming feature of his regimen was that I was the one delegated to work the same procedure on him. Dr. Stroud nicknamed me Dr. Heavy Hands. I called him Dr. Shroud.

The taciturn John Tolson, the same man who rushed to the prow of the *Southern Quest,* warning the public away as she crashed into Irongate Wharf, acted as film cameraman for the expedition. John represented the calm eye of the storm in our hurricane of conflicting personalities. He was like Edward "Uncle Bill" Wilson on Scott's expeditions: a man to whom all could turn. Everyone could talk to John. His silence was like a blank slate upon which we all scrawled our frustrations and fears.

The conditions kept us inside, locked in an abrasive state of hut fever. The winds that swept the beach clean of snow were a constant, howling menace. We fingered the kitchen knives and contemplated the backs of one another's necks.

At the end of our nine-month confinement in the hut, a scientist from the United States Antarctic Program, a woman named Elizabeth Holmes-Johnson, based at McMurdo, worked up a psychological report on the five of us. (Scientists in Antarctica have a lot of spare time on their hands to conduct random research projects.) In hindsight, the report, with its coolly objective scientific tone, makes for fairly entertaining reading.

"The individual need for autonomy," Holmes-Johnson wrote,

"overshadowed their desire to resolve their personal differences." In other words, we were all big babies.

Holmes-Johnson indicated that the others considered me "too involved in having the limelight" (big surprise there). I was thought "technically not competent to pull the group through the forthcoming event." As to who would make a better fit on our walk, Gareth was "individualistic," while Mike was judged "more stable."

Antarctica is a mirror. It throws back an image of yourself that you might not want to see. If you spend the winter there, it's quite a dark mirror. You tend to dwell on your thoughts.

"No man remains quite what he was when he recognizes himself," wrote Thomas Mann.

It turned out that a lot of things I believed I was—a cheerful guy, a boon companion, a natural-born leader—I actually wasn't. My psychological crutches became apparent. There are no strangers like married strangers, and I found myself at dinner around our big plywood table, thinking, *Who is this person next to me?*

Pete Malcolm had departed with the *Southern Quest,* leaving me, poor Robert, with a group of ultimate professionals. I was the youngest in the hut. I could feel that the others, at best, tolerated me. They had not enjoyed my incessant prancing around trying to get funding.

This was a head-down period for me. I knew I had to stay on the down-low, stay out of the limelight, and not try to assert my opinions in any way.

At the same time, in addition to all the negative stuff, I found aspects of myself that were not all that bad. During the ship-unloading process I hefted every single sack of coal. I didn't just lift one and have my picture taken. I did the whole job. Now, in the hut, I searched for a useful job—it had to be menial, to suit my menial skills. I became the hut's garbage person, flattening discarded containers, sorting out our trash.

But gradually I found myself smashing cans with a bit of extra violence. For my part, I became annoyed by Roger's gloomy temperament ("I am in a somber mood," reads an entry in his journal, "quiet and sullen"). Gareth's meticulousness and Mike's infuriating competence went into the mix, too. Jack Hayward Base turned into a pressure cooker. More than once, I thanked God for John's calming friendship.

When things got especially bad, I'd look over to Scott's forlorn hut at the other end of the beach and give myself a reality check: *He never made it back!*

It distressed me out of all proportion that I appeared to be the only person impressed by the history behind what we were doing. I felt alone in my obsession. I wanted to shout at the others: *Can't you see? There! There! Right down the beach—it's Captain Scott's hut!*

Roger wrote of me in his journal:

> He knows nothing of the reality that awaits us, only the heroic tales of History. Perhaps he can carry his History with him, it could sustain him, for he has in his mind no other model of the sanity of the action we take. When we step out on to the Barrier we will enter the unknown. No one knows if it is possible to make this journey unassisted, and we will not find out until it is too late to turn back.

Scott's Hut

Of course, Roger and the others could see Scott's 1911 hut just as well as I could. They just hadn't grounded themselves in Heroic Age minutiae while they should have been doing their university schoolwork.

As soon as we settled in, the books had come out, the blue two-volume set of *Scott's Last Expedition.* I immersed myself once again, although this time it was a very different experience, reading the journals while being able to look out the window to the very landscape Scott was describing.

Oh, right, I'd say to myself, *that's where Wilson painted that picture. That's where that photo was taken. That's what Scott was on about.*

Scott's hut itself was locked and off limits to casual visitors. It's a UNESCO World Heritage site, administered by the New Zealand Polar Division, which has done a bang-up job preserving it.

But Pete Malcolm's father had worked in the fabled MI6 (military intelligence, section 6), the secret-service arm of the British government. James Bond territory. Peter took his father's tutorial regarding the process by which any door could be opened and closed without anyone ever knowing it had been disturbed.

Pete could get through any lock. Before he left on the *Southern Quest,* he solved the door to Scott's hut. I thus had access to the holy of holies.

I was young. I apologize to the gods of world heritage for my trespass. My guilt forces me to send an occasional donation to UNESCO's World Heritage Site organization. UNESCO and the New Zealand government really do good work keeping the site safe from an invasion of ice and snow.

The first time I entered Scott's hut, I had the uncanny sense of having been there before, a result of my obsessive reading of journals and letters, not only of Scott himself but of Wilson and others who wintered there. I was home. The comfortable familiarity played off a countering sensation, a strange feeling of transgression, as though I was breaking into an occupied residence.

It was a very special, haunted place. I expected Scott-expedition stalwart Birdie Bowers to come around the corner and say, "Ah, Swan, we've been expecting you."

The dustless polar atmosphere left the interior pristine. A light rime, from the exhalations of visitors, covered every surface, but otherwise no one had dared to violate the spell cast by the surroundings. Here was a Dorian Gray of a place, the single room on earth, possibly, least changed by the passage of time.

Near the old stables lay a dog carcass from Scott's time, perfectly preserved. Outside the hut, we came upon a cache of butter and jam tins. Once thawed, they tasted fresh and more delicious than any butter and jam I have ever tasted from today's modern manufacturing process.

Shackleton's "lost men" of his Ross Sea party, marooned here in 1915, had removed the stacked wall of boxes between the officer's wardroom and the seamen's mess, opening the whole of the 50-by-25-foot hut to view. In every corner, items left by Scott and his men seventy-five years earlier still rested in place. Tins and bottles of Rowntree's Elect Cocoa, Bird's Baking Powder, and Lyle's Golden Syrup crowded pantry shelves near the cast-iron stove.

There were products from some of the same companies that I had sought out during fund-raising. Those efforts, even though they had gotten me there, now embarrassed me a little as I passed through the silent hut. They seemed as much a violation as my present intrusion.

I was drawn to Scott's den, amidships port. Of all the personal spaces in the hut, it had been most thoroughly cleared of items. Pinned to a nail was a fading picture of his wife, Kathleen, looking happy and fresh despite the curled edges and brown print of the photo. A memento of one of the world's great love stories.

"We took up our abode in the hut today and are simply overwhelmed by its comfort," reads the Tuesday, January 17, 1911, entry in Scott's journals.

While I steeped myself in history, Roger made repeated forays from

Jack Hayward Base, climbing twelve thousand feet up to stare into Mount Erebus's volcanic cone, heading south to visit the McMurdo bases, trekking across the interior of Ross Island to Cape Crozier.

Roger's energy, his double- and triple-testing our equipment, astonished me. I huddled in our hut as he left to embark on his adventures. My mates were men who thought it a fine lark to climb a mountain before breakfast. I was ill equipped to match them and terrified that I wouldn't measure up. I consoled myself with the allurements of the past, writing in my journal that "the history of this place is enough reason for being here."

Among the sixty-four tons of equipment we off-loaded from the *Southern Quest* were two mountain bikes, and to break loose from the clutches of the hut, we rode out onto the sea ice, a strange, desolate sensation.

As the winter wore on, anything *anyone* did began to annoy me, and everything *I* did annoyed me, too. I was crawling out of my skin, anxious to get started but terribly fearful of going. Claustrophobic and agoraphobic at the same time. A bad combination. There was no place for me.

I had tried over and over to psych myself into the proper frame of mind. No human being wants to die, I told myself, but I wasn't going to spend my entire time there in panic and terror. Roger, Mike, Gareth, John—each one could at least form a mental picture of himself walking to the pole. I couldn't. It was foreign to me. I tried to persuade myself that come November, I would leave the hut and reach the pole or not come back.

My hut mates eventually couldn't take my tapping. My letter typing annoyed them so much that I had been exiled up the beach to a makeshift shack located directly next to Scott's hut. I thrilled at the idea that I was slamming the keys only a thin wall away from where Scott himself wrote his journals and diaries.

I used onionskin paper, just as Scott did, and signed off some of my letters the same way, with "Your obedient servant." My correspondents must have concluded that I was very old-fashioned or quite simply daft.

Even though it was still months away, the time of our departure seemed to loom ominously before me. My nervousness and restlessness increased. I made a special visit to Scott's bunk, lay down, and stared up at the ceiling. Like the Little Engine That Could, I repeated, *I think I can I think I can.*

In the dead silence of Scott's ghost hut I muttered cajoling phrases to myself—and to him. "I will follow you all the way to the grave," I swore. Quite a lot of that was complete bull, of course, just a young man's gassy romanticism, but it did help to cheerlead myself forward.

To burn off energy, I took to jogging a circuit around Cape Evans in the winter dark, the cold air tearing great hacking coughs from my lungs. It took me thirty minutes to complete the three miles past such local landmarks as the Ramp, Skua Lake, Whale Cairn, Wind Vane Hill, and West Beach.

On one of these runs, Mike Stroud literally tripped over a stray emperor penguin. Both human and bird let out strangled shrieks. Emperors are the rare creature that winters over, but they were concentrated in a colony at Cape Crozier, on the other side of Ross Island. Mostly, Cape Evans in winter is bereft of life. The region's most reliable animal entertainment, in the form of Adélie, chinstrap, and gentoo penguins, decamped for greener seas.

Conditions were atrocious. The winds that poured off the Antarctic Plateau blew salty snow off the sea ice, and it clotted around the hut. At −39 degrees F the mercury column in a normal thermometer freezes solid. Roger discovered a bottle of Scotch in an abandoned hut on the flank of Mount Erebus, the liquor a block of amber-colored ice.

After four months of night at Cape Evans, a phrase from Henry Miller started to drum insistently in my mind. *We are all alone here and we are dead.*

South

His Grace returned on August 23, a pale flare poking up on the horizon from behind Mount Erebus. By late September, I could put my face to the sun and actually feel warmth. The oppression of the polar night becomes most apparent as it lifts. I wondered how I'd made it through. I almost hadn't.

With the sun, lightness and humor and possibility returned to the world. Adélie penguins visited the beach in front of the hut, constantly curious about what we were up to. It is almost impossible not to laugh out loud upon being approached by a penguin on land, its waddling, fat-man gait is so comical. But penguins are out of their true element on land. In the water they transform into beautifully sleek missiles. Human beings must look risibly ridiculous, too, plumped up in space suits and waddling across the surface of the moon.

Our days that spring were consumed by preparations for the walk. Mike and John did most of the cooking and hut work, leaving me, Roger, and Gareth free to get ready. Every ounce of weight had to be stripped away from our kit. I saved a few grams by whittling down the handle of my toothbrush. We spent a painstaking couple of hours stripping the wrappers from 480 Yorkie candy bars (made by Rowntree's of York, "where the men are hunky and the chocolate's chunky"). Total weight savings: twenty-six ounces.

Our sledges were molded Kevlar built by Gaybo Ltd. of East Sussex, equipped with what were, back then, state-of-the-art Fluon runners. The sledges were fitted with distance wheels from Trumeter, which would click off our progress for the nine hundred miles to the pole.

Sledge, equipment, provisions, food, and fuel totaled 354 pounds for each man. From that weight we had to obtain the calories necessary to fuel our walk. It was a delicate equation. If the sledges were too heavy, too loaded down with extra provisions, we would burn more energy over the long run than we could comfortably haul. Ideally, we would arrive at the South Pole with near-to-zero calories left in the supplies.

Conceivably, though, we'd run through all our calories before we got there. What then? There was no such thing as running on empty in the polar environment. The human body uses up calories at an incredible rate just in the struggle for existence in such harsh conditions, never mind sledge hauling.

One piece of equipment we didn't include in our kit was radios. On the face of it, this was a crucial decision, because it changed the whole color of the walk. Without radios, we truly would be unassisted, unable to summon rescue. Radios meant extra weight. I think the real reason we left them behind, though, was that they would somehow adulterate the purity of our purpose. We wanted to be isolated, we wanted to be alone, we wanted to respect Scott's conditions as closely as we could. Going without radios would be one way to accomplish that.

Unassisted. There are a lot of fantastic treks nowadays, adventurers taking off for the outback wearing nothing but sandals, walking the equator backward, stunts such as that. But I blessed Roger for his insistence on no assistance. It forever set In the Footsteps of Scott apart. It meant that once we vaulted out into the unknown, we were left to survive solely on our terms. The salient point about an unassisted march is that once you pass the halfway point, anyone injured must be left to die, in order to save the remaining expedition members from starvation if they attempted to lug the injured party to safety.

Our sledges were vaguely coffin-shaped. Kevlar sledges are used most often today at ski resorts, to rescue injured skiers. On the return from the southern journey on the *Discovery* expedition in 1903, Shackleton was so debilitated by scurvy that he had to be sledged by Scott and Wilson. I had a disconcerting vision of myself fitted neatly into my Kevlar sledge, arms folded across my chest, my dead body hauled by my friends. I tried not to take myself so damned seriously, but I couldn't help but dramatize the occasion.

The decision of who to take seemed to become more fraught each day. It was a ridiculous error to have left it hanging fire so late in the process. Mike or Gareth? Gareth or Mike? Logistics or safety? An obsessive-compulsive or a genial physician?

Roger backed Gareth, while I held out for Mike. Slowly, Roger's superior authority in mountaineering matters wore me down. In August I wrote in my diary, "I think I am doomed to GW." By the Antarctic spring I had come around to honoring Roger's judgment.

Mike accepted the decision with a resigned sense of bonhomie. Gareth it would be.

"The weather looks grimish but we have to go," I wrote in my journal. "I feel nervous but now I want to complete the job. I leave in fine company on a great journey."

On October 25, 1985, we departed from the hut with our packed sledges, headed for our departure base camp on the barrier. John Tolson came along to film, and Mike Stroud hauled extra supplies. We began by trudging through a ground blizzard, battered by frozen pellets of spindrift blowing in off the sound.

One of Roland Huntford's primary criticisms, in his book-length attack against Scott, regarded using sledges to "man-haul" supplies, rather than dog sleds. Huntford rigged his compare-and-contrast portraits of Scott and Amundsen so that the latter appeared well reasoned and the former deluded.

The Norwegians considered the man-hauling harness "an instrument of torture," Huntford reported, and they could not understand why anyone would decline the use of dogs.

From a historical point of view, I thought otherwise, and Roger Mear worked out a convincing rationale for the efficiency of man-hauling. The comparison was not at all as clear-cut as Huntford contended. Nansen, Amundsen's mentor, man-hauled on his trek across Greenland. Dog teams worked out to be superior only if one used the expired animals, which were inevitable, as meat, for the other living dogs and for humans, as Amundsen did. PETA would be appalled. (The coldly logical Amundsen has many passages in his memoir, *The South Pole,* professing both love for his sled dogs and a relish for their meat: "The thought of the fresh dog cutlets that awaited us when we got to the top [of the Axel Heiberg Glacier] made our mouths water." The cutlets in question belonged to the dogs making his ascent of the glacier possible.)

Our equipment was leagues ahead of Scott's. We had Ramer Grand

Tour skis, Therm-a-Rest air mattresses, down sleeping bags and outer-sleeve bags from Mountain Equipment, Beal glacier ropes, Chouinard ice axes, and, yes, three Victorinox Swiss Army knives. Working with Roger, the Mountain Equipment folks developed innovative vapor-barrier down clothing. Now standard polar technology, the gear made sure our sweat did not turn to ice inside our clothes—a constant complaint of Scott's team.

As for provisions, we would survive for eighty days on such delicacies as military service biscuits. Major Cordingley had dispatched one of his lieutenants, Mike Hough, on board the *Southern Quest* with more than three thousand of the bland but nourishing biscuits, a staple of army fare. We also packed Cadbury's instant hot chocolate (eight pounds), Maggi instant soup, Raven freeze-dried peas, Mountain House bacon bars (twenty-six pounds), and pepperoni (forty-seven pounds).

Mountaineers in those days discovered that oleomargarine gave the best bang for the buck, or rather, the best calories-per-ounce ratio. There were stories of Denali expeditions surviving on cases of the stuff, eaten stick by stick. The resulting halitosis was supposed to be tremendous.

We weren't quite that stripped down. We melted half sticks of tinned butter in our soups. Much of our provisions were slipped to us under the table from the mess of the Fifth Inniskilling Dragoons. We even allowed ourselves one four-ounce stick of Mitchum's antiperspirant each, not for civilization's sake so much as to keep our feet from sweating, since sweat turned to ice in the conditions we would face.

Thus supplied, provisioned, and burdened, we set out from the edge of McMurdo's Williams Field, out on the barrier seven miles from the big American base station. The skiway, universally referred to as "Willies," is the busiest airfield on the whole continent. The supply Hercs landed there, and anyone flying to Amundsen-Scott Station at the South Pole stopped at Willies first. The flight from McMurdo to the pole took three hours.

We were going to take a longer way around, to say the least. Mike and John would accompany us for the initial few days, filming our progress. About two dozen friends from the McMurdo and Scott bases gathered to see us off. The Kiwis presented us with a bon voyage bottle of Scotch. We left it with John and Mike because we could not afford to carry the extra weight.

November 3, 1985. The anniversary of Scott's first camp on the barrier. "The future is in the lap of the gods," Scott wrote in his journal. "I can think of nothing left undone to deserve success."

I slipped on the Troll-brand sledge harness that would be my home for the next seventy days. "With a couple of jerks I was off" was the joke we told one another later. I surged forward, sledging a load behind me that was double my body weight. I found the going extremely difficult, but I reasoned that the load was now the heaviest it would be on the whole trek.

This was the weight I would haul almost nine hundred miles, consuming an identical amount of calories in the effort as was contained on the sledge. I would haul the load over hard-frozen black ice, as well as blue or glacial ice, white ice, a.k.a. snow ice, névé or refrozen snow, graupel, powder, firn, corn, boilerplate surface hoar, soft crust, glazed crust, ice knobs, snow hummocks, wind slabs, ground blizzards, pillow drifts, jumbled glacial blocks—all of the myriad forms of ice and snow we would encounter on our path to the pole.

My sledge and I would also have to cross thousands of ice crevasses, treacherous and often camouflaged by smooth, deceptive surfaces. The weight I hauled made me all the more nervous about them. I had somewhat wildly hoped to avoid them altogether. Instead, we encountered them immediately.

The Great Ice Barrier

Crevasses are the killers of the Antarctic. "A grim trap for the unwary," Scott called them. Glacial ice tends to crack and fissure, creating narrow, jagged gaps in the surface. Oftentimes snow will then blow over the gap, until the surface presents a smooth, deceptive face. Underneath the crust lurk fifty- or one-hundred-foot drops into icy ravines.

The Great Ice Barrier acts like a slow-motion sea. It appears flat and static, an immense lozenge of ice. But fed by the glaciers that extend off the Antarctic Plateau in the center of the continent, the ice of the barrier is in the midst of constant, slow-grinding flow. Wherever humps of land extrude into that flow, wherever a glacier meets terra firma, the obstruction tends to create ripple upon ripple of deadly, invisible crevasses.

Of the five men in Scott's South Pole team, the first to die lost his life in a crevasse fall. Edgar "Taff" Evans, a Welsh petty officer who was the sole noncommissioned officer on the trek, was returning from the pole, crossing what Scott characterized as a "good hard surface," when he and Scott both tumbled into a crack in the ice.

Later commentators have concluded that Evans suffered a concussion in the accident. Scott writes that the burly Welshman, normally "a tower of strength," became "dull and incapable" from that point on. Two weeks after the crevasse fall, Scott noted: "Evans nearly broken down in the brain." The following day he was dead.

For anyone walking on the barrier, the ice crevasse becomes a constant bugaboo, a chimera, a memento mori, haunting and haunted. Staring down into one is like looking into a fresh grave. You can't prevent yourself from turning to philosophy. "Come lovely and soothing death," as Whitman wrote.

It doesn't help that ice crevasses at times display great beauty. The crevasse, a nineteenth-century European glaciologist wrote, "gradually widens until it becomes a broad chasm, shimmering in the sunshine with a blue light, delicately pale near the surface of the glacier, but deepening to a dark Prussian blue or indigo in the depths."

I had a more immediate reason to be terrified of crevasses. One of our mates at Rothera, an experienced mountaineer named John Anderson, had perished with another man in a crevasse fall. If even a top veteran climber such as Anderson could die, I thought, what chance had a neophyte like myself? And with no radios, my worries over an accident were exacerbated.

I thought about John Anderson and Taff Evans as we pulled away from Willies Field at the start of our journey. I carried with me Evans's polar medal, stitched into the lining of my jacket, a gift from my brother-in-law, John Drew. Scott commented on the irony of the strongest among them—the stocky, six-foot-plus Evans—being the first to weaken.

I couldn't shake the nagging feeling that I would be the Evans of our expedition. I was the youngest and, thanks to the twin legacies of the Sedbergh run and of tree hauling, the strongest. Would I also be the first to falter?

To the south, White Island's Cape Spencer-Smith (memorializing another Heroic Age death on the barrier) represented the kind of land obstruction that would likely cause crevasses to form in the ice. Roger, in the lead, gave it wide berth, heading almost directly east, into the great expanse of the barrier itself.

In planning the expedition, we estimated that we would have to cross some six thousand ice crevasses to get to the South Pole. We agreed that if we stopped and debated over each crossing, we would lose precious time. Our food and supplies were gauged to the ounce. We couldn't afford to dawdle.

But at our first verified encounter with a crevasse, we stopped and debated and dawdled. Roger halted to allow Gareth and me to huff up to him. He pointed with his ski pole: the broken ice crust, darkness beneath.

"Well, they're usually lateral—" Gareth began, pointing his ski pole forward.

"Unless they're transverse," Roger interrupted.

I studied the almost featureless surface of the snow. Could I detect a faint ghost line running across our path? Was this it? So soon?

After more bickering and pointing, Roger bulled ahead. We were on skis; we hauled our sledges across our first crevasse without incident.

Our lack of team cohesion early on did not bode well for the 900-mile trek in front of us. The journey broke down into three stages: barrier, glacier, plateau. The Great Ice Barrier, a.k.a. the Ross Ice Shelf, accounted for 400 miles. The Beardmore Glacier, only 120 miles, but with an elevation gain of 9,000 feet. The Antarctic Plateau would cover the last 350 miles. On the ice shelf—the ice cube—we were barely 1,000 feet above sea level. The Beardmore was our ladder to climb onto the plateau. The plateau itself, vast, high, and as featureless as the barrier, held our prize, the pole.

It took a while, over those first few days, to sort ourselves out. The immensity of the barrier flattened us. We stood on a frozen pan of ice the size of the American Deep South—Florida, Georgia, Alabama, Mississippi, and Louisiana put together. No magnolia-scented breezes here, though. The monotony of the horizon seemed to work like a lid pressing down on my skull, crushing my spirits.

The tent we erected each night and struck each morning (though in the Antarctic summertime light, night and morning were merely figures on a clock) turned out to be a tight fit. I slept on the left, my head toward the entrance, while Gareth, the tallest, lay the opposite way in the middle. Roger was always on the right, his head next to Gareth's.

Every morning I was the last to dress and the first out of the tent. While Roger and Gareth worked the stove and prepared the flasks of soup for the day, my self-appointed task was to bring in chunks of snow from just outside the tent to melt for water. I thus got the first peek at the weather.

The fare varied little. Oatmeal, soup charged with corn oil for added calories, chocolate, cold slices of pepperoni. But I learned quickly that deliciousness is a relative quality. Gourmands might turn up their noses at our meals, but my ravenous hunger rendered them five-star, beyond five-star. A simple square of chocolate became manna of the gods. It didn't matter that I had eaten an identical square only hours earlier. To my lips and tongue, it was always new, always a revelation. In the gruelling routine of my day, in the flat, featureless skillet of the barrier, chocolate melting in my mouth represented a welcome break in the monotony, a bump in the boredom, an *event*. Heaven, bliss.

Lacking any reference point (horizon, same; sky, same; surface, same), I fell back on history to inform my days and ways. I cut loose from time. I was not here and now, I was there and then, with Scott, Shackleton, and Amundsen on the barrier.

Incidents straight out of polar journals haunted me at every step as I marched through the past. Here was where poor Shackleton had to be loaded onto the sledge in Scott's first southern journey in 1903. He heard Wilson tell Scott outside the tent that Shackleton would die before they completed the trek.

"I'll outlive you both," Shackleton told them. Indeed he did, when Scott and Wilson perished in 1912, very near the place on the barrier where they had declared Shackleton a dead man ten years before.

Every point in the barrier seemed alive with incident for me. I knew as I passed the exact spot where, three quarters of a century earlier, the skittish demon pony Christopher tried to bolt, but Captain Titus Oates (he of Patrick Cordingley's Fifth Inniskilling Dragoons), the expedition's horse expert, held on.

And just here was where the Scott expedition arrived at Bluff Depot, after "uniformly horrid" marches over "wretched" surfaces.

> The snow which had fallen in the day [Scott wrote] remained soft and flocculent on the surface. Added to this we entered on an area of soft crust between a few scattered hard sastrugi. In pits between these in places the snow lay in sandy heaps. A worse set of conditions for the ponies could scarcely be imagined.

Yes! I had read this passage repeatedly, but always in the safe confines of England. Now I knew what Scott's words meant in ways a mere reader never could. Now I knew how snow could be described as "sandy," when the weather was cold enough to render it glideless beneath the runners of a sledge.

Sastrugi—the word is Russian in origin—were something I had read about over and over in polar literature. But here they were in fact: rock-hard, yard-high frozen ridges of snow, splayed this way and that in fanciful, windblown patterns, usually directly across one's path. Constantly climbing over them was like being put through a commando-training obstacle course while hauling a 350-pound sledge.

The philosopher Bertrand Russell first suggested the distinction between knowledge gained by description and knowledge gained by acquaintance. I had acquired all mine by description. Now I was making its direct acquaintance. Humping over those damned sastrugi, I didn't know but that I rather preferred a comfortable chair, a warm hearth fire, and some well-wrought English-prose description.

The Great Ice Barrier—I always referred to it like that, never by its more proper scientific name, the Ross Ice Shelf—seemed an obstacle to get across, a wall to climb over, a struggle to endure. A large part of the barrier's awfulness lay in its lack of features. White, white, white, no color. Contrast, zero. All points of the horizon identical. It offered no landmark by which to orient oneself, and worse, no goal to which to aspire. *If I can only make it to that mountain* . . . But there was no mountain. There was nothing.

"Far was as far as near, and near was as far as far" was how Roger put it.

Early on, for me, there was only the sledge, the harness. I tried to make it my friend. I saw no other option. But how do you not grow to hate your torturer? Always it was there, the weight, the pull, the dull slog. Slide one ski forward and pull. Now slide the other ski forward and pull. Repeat ad nauseam.

For a while I took a mantra from Samuel Beckett: "I can't go on, I can't go on, I'll go on." I got into a little rhythm with it. *I can't go on*—pull. *I can't go on*—pull. *I'll go on*—pull. Then I took a line of Robert Frost: "The best way out"—pull!—"is always through"—pull!

But words palled on me after only a short time. My fascination with history faltered. Stopping often to take mouthfuls of snow, I was continually consumed by a terrible thirst. I began to hallucinate shapes and patterns in the whiteness around me. My mind played tricks, wandering in fanciful directions, always getting thudded back down to earth by my friend-enemy, the sledge. I was pulling a great oaken log through the sand.

Nothing worked. Falling farther and farther behind Roger and Gareth, I began to despair. I was dead weight. They were the leaders, my guides. Roger and Gareth were the ones who belonged here. I did not. I was indeed what they had secretly labeled me among themselves: "the passenger."

I can't explain how hateful this feeling was to me. For the first time in my life, I wanted out. The sense of failing hurt me, eventually becoming a worse drag on my progress than the physical torture of the harness.

Then, quite by accident, Roger discovered the flaw that had been holding me back.

Runners

Somewhere on the barrier, encased beneath a sarcophagus of ice, the bodies of Scott, Bowers, and Wilson remain entombed in their tent. Given rates of snowfall and windblown blizzards on the surface, they are probably a few dozen feet below the surface by now. Because the ice shelf itself moves, they are fifty miles or more from the actual compass point where they died.

Two weeks and a little more than 150 miles into our trek, we passed that compass point. Twenty miles farther on, we arrived at the place where Titus Oates died. Nearly crippled, slowing down Scott's desperate but ultimately unsuccessful run to safety, Oates sacrificed himself for the good of the others.

Scott's journal recorded Oates's farewell words: "I am just going outside and may be some time." The diffident, self-effacing tone somehow spoke to the national character of Britain, and the phrase was repeatedly called out and quoted when the journals were published in the aftermath of the disaster.

We plunged onward, struggling always, approaching the place on the Beardmore where the Scott expedition's first human casualty, the crevasse-stunned Taff Evans, had been buried over seven decades before.

It was there, at the foot of the Beardmore, as if to honor the death of the man whose posthumous polar medal I carried in my breast pocket, that I realized our expedition was breaking down.

I myself was breaking down, in extreme physical agony over the difficulty of pulling the sledge. And relations between the three of us were disintegrating to the point of no return. Roger and Gareth's petty bickering turned acid.

"You're too intolerant," Gareth said to Roger. "I feel like I can't say anything because you think whatever I say is such a bore."

"Constant complaining and moaning *is* a bore, you're right,"

Roger said. *Read your diary,* he told Gareth. The real truth was likely to be written down where you think no one else will read it. "Your fears so dominate your thoughts that you're unable to concentrate on what needs to be done so those fears don't materialize."

We had no GPS, no satellite tracking. Roger and Gareth navigated using a sextant and a watch. We kept on course by compass, and Gareth, at least, was having difficulty whenever he was in the lead. His course tended to wander, and we could simply not afford the extra steps.

Gareth's feet became blistered. He was, I believed, in the worst shape of the three of us. But we each had problems. Roger's moods were darker than ever. Though I habitually exclaimed to the others how easy it all was, in reality I became more and more convinced that I would not be able to complete the trek.

I endured. I kept on. With Scott's death-tent site behind us, we approached the "Gateway," the Beardmore Glacier. The ridges and peaks of the Queen Alexandra Range poked up on the horizon to the south, an immense relief. Finally, a goal to work toward. But the work continued to be agonizing.

On December 4, a month after we started, Gareth stopped us again with foot trouble. We had come to the very edge of the barrier. More important, we were approaching the literal, not figurative, point of no return. From there, it would be just as far to return to McMurdo as to go on to the pole.

Up to the halfway point, if someone got injured, we could conceivably chuck away some equipment and food and put whoever went down on a sled. Take him back home to McMurdo that way. But after the halfway point, if someone was unfit to continue, we would have no choice but to leave him behind. To jettison food at that stage would mean the deaths of the uninjured two as well.

I realized I had to make what was, for me, a terrible, terrible decision. I wandered away from the tent soon after we pitched it. That was when I made my pathetic, weepy "just don't kill us" plea to the continent that seemed intent on doing exactly that. What would happen? How would we get out of there? Every option appeared impossible.

I chose door number one. I decided to quit.

I headed back to camp. "Listen," I said to Roger and Gareth. "I've got something to say that will blow your heads off."

Annoyed that my dramatic announcement did not prevent Gareth from picking at his blisters, working to keep his feet in shape, I nonetheless continued. "I'm worried that you won't make it," I said to Gareth. "I've dreamt about this journey for years. But I just walked out onto the barrier and I dug down deep in myself. I realize that making the pole isn't everything. I would feel awful if someone died because of my stupid ambition."

I paused dramatically. Gareth kept up his foot hygiene. "I think we should consider turning back while we can and leaving Roger to continue to the pole."

We got into it then. Roger agreed we couldn't afford delays. Gareth didn't understand what the fuss was about.

Roger said, "If a helicopter landed right here, right now, Gareth, would you get on it?"

"No way," Gareth said. "You bastards would have to put me on it. I want this as much as you."

I felt like a fake. Gareth got all the heat, while I knew in my heart of hearts that I could not keep pulling my sledge for another four hundred miles, up the Beardmore and across the plateau. Even the next four hundred yards was impossible. I simply could not do it. My dream had died.

I can't go on, I can't go on, I'll go on—well, no, actually, I really, really can't go on.

The Beardmore winds blew my promise back at me. *Just don't kill us, and I promise I will somehow do whatever I can do to protect you.* Uh-huh. Big words, chap. You're in no shape to do protecting for anyone. You hardly know how to protect yourself.

During our heart-to-heart, I gifted Gareth with a piece of my sausage. A small gesture, but one he still refers to, years later, whenever we get together. Within the Darwinian world we had created for ourselves, sausage giving was a supreme act of love.

Somehow, the gauntlet I had tossed down wasn't taken up. We

talked far into the night, sorting out grievances. And the next morning we got up, ate our breakfasts of oatmeal and margarine, and strapped on our harnesses.

Agony. I fell behind almost immediately. After three hours, the mountains appeared to be receding rather than getting closer. Roger and Gareth were so far ahead I could not even see them, provided I would ever lift my eyes from the tips of my skis, which I did not.

Roger and Gareth took our traditional midmorning break, and Roger skied back to help me. He hadn't realized how far I had fallen behind—a full mile. I managed a grin as he skied up, and we pulled together for a while. Finally, Roger suggested he should take the sledge alone and I should ski forward and join Gareth.

As soon as Roger pulled alone, he realized that something was terribly wrong. Even lightened of a month's supplies, my sledge pulled much harder than his. He couldn't believe it. Why hadn't I complained? (I was tough!) Did I have a muscle for a brain? (Yes!)

As soon as he pulled up to Gareth and me, he suggested we put up the tent and call it quits for the day. I looked at him strangely. It wasn't like him. We hadn't made our quota of hauling hours—a nonflexible quota of nine hours a day, seven days a week, seventy days in a row, only abandoned if it was blowing too hard.

"I think I have a surprise for you," Roger said. He had me empty my sledge, and we hauled it inside the tent. "Look at this," he said, passing his hand over one of the runners. The state-of-the-art Fluon surface wasn't supposed to have a grain, but it did. My runners had been inadvertently attached backward. I had been pulling against the grain the whole time.

I sat, stunned, as Roger dismantled the runners, drilled new holes (using only a Swiss Army knife), and reassembled the whole rig. Checking Gareth's sled, we discovered one of its runners was also attached backward, leading the unit to skew in one direction. Gareth's back muscles had felt continually out of kilter, and now the reason for it was revealed.

In the overheated political dynamic of our little trio, Gareth and I even shared the suspicion that Roger had intentionally reversed our runners. But deep in our hearts we knew it to be a simple mistake. A simple mistake that almost broke our backs.

The next day was like magic. I floated. The sledge runners glided smoothly where formerly they dragged. I couldn't believe it. The comparison would have been like night and day, had there actually been a night or day in our present environs. I cursed myself for my foolish stoicism.

Even though I followed in the footsteps of Scott, I would never compare what we did to what he did. The brand of greatness that Scott, Amundsen, and Shackleton achieved isn't available to us in this day and age. They accomplished extraordinary, almost superhuman heroic treks, and they did it without freeze-dried food or ripstop nylon or the myriad other benefits that the modern age bestowed upon me.

But though I can in no way place myself in the same class as Scott, I continually sought out the parallels between his life and mine. And one astonishing parallel still lay before me, up smashed-glass ladders of the Beardmore and across the Antarctic Plateau, as we proceeded to the South Pole. It was this: In his moment of greatest triumph, he felt, as I did, the horrible sting of great catastrophe.

Ninety Degrees South

Almost without our being aware of it, we became a team. One of the first achievements in that direction came soon after we started crossing ice crevasses. Before we embarked on our trek we had agreed not to dawdle or debate when negotiating these, but we had left undecided the matter of *how* to avoid debate and *how* to cross each individual crevasse.

We knew that forming a focus group and plotting a course wasn't the way to go. We'd still be there today if we had done that. After a half dozen stops and starts and useless bickering back on the barrier, we developed a method that served us well. The attack was simple: Whichever one of us approached the crevasse first decided the way across. Invariably it was Roger. No discussion, no debate.

That became an unalterable rule on our trek. It helped us cross those thousands of crevasses without a serious fall. More important, it forced us to trust one another. The practice forged us into a team and it led directly to our eventual success.

I have found this no-debate device very useful, even in team building that doesn't involve making life-or-death decisions. Someone makes the call; the others on the team trust it. Yes, there are some cases when a focus group is just the thing, but a lot of other instances benefit from the ice-crevasse principle we forged during In the Footsteps of Scott.

Trust is a squirrelly human quality, tied up in issues of control that speak deeply to our quite natural individual concerns over personal security and safety. It's easy to say that you trust someone. But it is very difficult to actually do so, and do it from the heart.

Part of my trust in Gareth and Roger stemmed from pure laziness. It was easier to trust them to get me to the pole than it was for me to learn the complex principles of navigation at the bottom of the world, where the compass needle marked south by pointing north, toward the southern magnetic pole.

I can tie exactly two knots, a figure-eight and a bowline. I have no real technical understanding of camp basics like stove operation and fuel economy. For limited distances, I am able to follow a straight line. That's about it.

Roger and Gareth could not believe I was willing simply to leave it all up to them. "Rob, what if an ice crevasse eats both me and Gareth?" Roger asked me. "What would you do? You can't navigate. There's no radio to call for help. What the hell would you do?"

I'd be lost, I knew that much. There's no way a human being can walk a straight line in the featureless polar wastes. One leg is always stronger than the other, always favored, so eventually the hapless wanderer will find himself circling around to cross his own trail.

I pictured myself alone in the Antarctica wilderness. I still dream about it today, over two decades later—being lost, a solitary figure in an icy immensity, without hope or prospect of rescue. No direction home, as Bob Dylan sings. It's always a nightmare of a dream.

"Tell you what," Roger said. "If we both fall into a crevasse, the only thing you can do is jump in after us."

I gave him a halfhearted chuckle. I knew he was right.

Along with the importance of trust, there was another secret of leadership I took from Roger. There exists a truism that mountaineers tell one another. Everyone wants to *have climbed* the mountain, they say, but very few people actually want to climb it.

A lot of us might desire to be able to say we've climbed Eiger in the winter. That has to be among the coolest claims a human can make, enough to impress even a football star, a guitar god, or a president. But actually to climb Eiger in the winter? Um, no thanks. As long as it's in the past tense, it's great. If we're facing the pain and the struggle of the task to come, not so much.

It's true for mountains, and polar expeditions, and a lot of other monumental tasks. Everyone wants to have sailed across the Atlantic; nobody wants to do it. Everyone wants to have written a novel; nobody wants to sit down and crank it out. Everyone wants to have achieved greatness, but few among us want to do what it takes to become great.

Roger was one of those rare birds who actually relished the doing, rather than the having done. He was a genius at doing. He liked the

struggle, looked forward to obstacles, persevered through pain. Oftentimes he infuriated me, but I learned an immense amount from him. He
was the hard man among us.

There comes a time, if you are going to embrace a leadership role,
when you have to execute. You have to do what you say you're going
to do. I relearned that lesson from Roger. I said I was going to walk in
the footsteps of Scott to the South Pole. If I had returned home without
doing so, what chance would I have of getting people to believe in me
again?

The secret is to find your purpose in the actual doing, not in the re-
ward that comes from having done. Find enjoyment along the way, in
the simple square of chocolate, for example, melting in your mouth on
the Great Ice Barrier. Enjoy the company of your teammates. All easier
said than done, of course, but all essential to a successful team effort.

Part of leadership involves a leap of faith. If I'm going to follow
you, I have to be able to trust that you'll be able to execute the task at
hand. In order to inspire trust in others, you need to trust yourself first.

In my own small, limited way, I performed a leap of faith trusting
that Roger could lead me to the pole. He was a mountaineer with a lot
of experience, so my trusting him wasn't that extreme. He performed
a much larger leap believing that I could mount a successful polar
expedition. I was a novice with only passion and energy to offer, so
his belief was more unlikely. At times I thought it nothing short of
miraculous.

Team cohesion and trust got us up the fearsome Beardmore Glacier. Like a candle in the wind, the dream that started while I read
Scott's journals in Durham or even before, while I sat in the dark at age
eleven watching John Mills in *Scott of the Antarctic*, somehow stayed
alive. I had sustained it over the years.

On Sunday, December 22, seven weeks to the day after we had
started our journey, I quoted John Mills as Robert Scott when we stood
atop the Beardmore. "Barrier, done. Glacier, done. Plateau ahead."

We turned south. We marched nine hours a day. Our daily miles increased, to 15.57, 17.00, 17.28, 17.22, 16.51. We averaged above 1.5
miles per hour always, and on some good days we approached 2.

We paused to honor Shackleton's decision, on his "farthest south" trek, to save himself and his expedition mates by turning back when he was only ninety-seven miles from the pole. On New Year's Day 1986, we hit three-and-a-half-foot-high sastrugi and threaded our way through them, still managing to tick off 15.57 miles on the sledge meter.

I had not allowed myself to think about the pole. It was enough to concentrate on what was going on in front of the tips of my skis every day. But now, like the light of the polar sun breaking the horizon after months of darkness, a tiny ray of thought appeared in my mind. As always, I thought about Scott, what he felt at this point. He must have known he was going to make it.

On Saturday, January 11, an object broke on the horizon for real. We had been stopped by a blizzard less than ten miles from our ultimate goal. The winds that day were the strongest recorded at the pole since 1958. We started again only after the storm died out, leaving the atmosphere silver with suspended ice crystals. I saw it to the west: a small, blocky structure poking up from the unremitting flatness of the plateau.

"There!" I yelled to Roger and Gareth as I waved my ski poles.

The structure housed weather instruments. It appeared odd and out of place in the barren expanse we traveled. A line of green flags led from it to the Amundsen-Scott Station at the South Pole.

Roger and Gareth both told me later that they'd sensed the same thing: The closer I got to the pole, the more I seemed to grow, they said, getting larger because I was so pumped with excitement.

I dislike our rote following of the flag line. I wanted to lose all constraint and leave behind the caution by which we had lived for so long. Part of it was the feeling of finally grasping the dream, part of it was triumph, and some of it was a mare-to-stable rush for shelter.

More surprising to me was the tinge of sadness I felt. I was sorry to see it end.

The Americans housed the Amundsen-Scott base beneath a huge geodesic dome, which shone on the horizon as we powered forward, an Emerald City in silver. I halted briefly to prepare my kit for arrival. I took out Taff Evans's polar medal from its home in the lining of my jacket. I wrapped a scarf Lavinia had given me around my head. I propped up her gingham Teddy in my pocket. Thus accoutered, I went forward humbly, head bent for the laurels.

Only it didn't work out that way. No one saw us until we were almost at the huge truck-bay entrance of the base. A human figure obscured by down jacket and fur-fringed hood stopped short upon seeing us. We weren't expected for another week—Scott's schedule would have had us at the pole on the seventeenth.

The pressure of reality distorted my perception. After the long monotony of the plateau, here were events, all at once, a spill of objects, colors, people! "Footsteps are here!" I heard called out as we entered the technological cathedral of the dome. "Footsteps! Footsteps! Footsteps!" I felt I was moving very slowly in relation to other people, who appeared to me like figures on speeded-up film.

Something was wrong. I could not figure out why people were crying. It didn't fit with the hail-the-conquering-hero reception I was going for. Was the relief at our survival so extreme that it moved people to tears?

Lee Schoen, officer in charge of the base, came forward to welcome us. "I'm glad you guys made it," he said. "But I've got some bad news for you. The *Southern Quest* is sunk, crushed by pack ice. Everyone is safe and being evacuated to McMurdo."

Psychic Emma had proved correct. We reached the South Pole at 11:53 on January 11, 1986. Three minutes earlier, the expedition's ship had disappeared beneath the polar sea waters off Beaufort Island, its hull ruptured at the engine room.

Just then is when it all went mad.

Madness

In very short order, via a series of airplanes, Roger and I found ourselves standing on the tarmac at Heathrow Airport in London. We were a bit thin and more than a bit worn. We hadn't taken a bath or shower for a year, which is extreme even for an Englishman.

A reporter informed us that we immediately had to go on camera live for a TV news show.

I've seen the footage. We appear haggard, blown out. Less than forty-eight hours earlier, we had stood at the South Pole and found out that the *Southern Quest* had gone down. I was acutely aware that ten million people were laughing at us. Everyone was laughing. We looked a bit strange, like just-landed aliens, which in effect we were.

So what was the interviewer's first question? What was the world itching to know after three daft Brits spent seventy days having the worst possible time for a totally pointless reason?

"Was it cold?"

At that moment, I felt a great blast of relief, since I knew there was real hope that I might not be the stupidest bastard on earth.

While sitting in a comfortable chair under the bright TV lights, with everyone in the country laughing at me, I thought about madness.

Several times over the course of organizing In the Footsteps of Scott, I heard people repeat the same two lines to me. It happened so often that I began to think of humankind as a species of oddly shaped parrots, everyone squawking out the same words.

They customarily say the first line with a big smile: "Rob, you're definitely going to fail."

Then, with an even bigger smile still: "You are definitely going to die."

When I returned home to so-called civilization, after I had failed to fail at the one thing that would, no doubt, according to the parrot people, kill me and all my expedition mates, the same people welcomed me back with a twisted smile.

"Rob, there's only one reason why you did this," they said. "You are completely insane."

I hope no one reading this has actually starved. Or ever will starve. I have. Lots of times. Voluntarily. When you starve, your body eats itself. You never forget that sensation.

So with the TV lights shining in my eyes, and everyone saying that I was mad, crazy, insane, and that it was all so funny, this is what I thought: *Half the world starves every day. The rest of us do everything we can to try to lose weight.*

Was that insane? Was that not a bit crazy?

We use fossil fuels at such a rate that they will soon run out, but not before they render the only home humankind will ever know into an uninhabitable desert.

Is that not lunacy?

Mark Twain: "When we remember we are all mad, the mysteries disappear and life stands explained."

The *Southern Quest* sank because her hull caved in under thousands of pounds of pressure exerted by freezing ice. She thus joined the list of Antarctic shipwrecks, a surprisingly short list given the lethal seas in the neighborhood.

There are six in all: famous ones, like Shackleton's *Endurance* and Swedish geologist Otto Nordenskjöld's *Antarctic*. A West German tanker, the *Gottland II,* and an Argentinean ship. More recently, the cruise ship MS *Explorer* struck an iceberg and sank close to the South Shetland Islands in the Antarctic Ocean.

And the *Southern Quest.* Her addition to the short list was a dubious achievement at best. Captain Graham Phippen maneuvered the former trawler within fifty miles of the Jack Hayward Base on Cape Evans. He then turned her north, where pack ice and icebergs found her near Beaufort Island. As Phippen nosed through open-water leads, they closed around the ship. One starboard iceberg in particular, acres huge, rotated into the ship's side. The bulkhead in the engine room bowed inward and then burst.

In a larger sense, the *Southern Quest* was crushed not by polar pack ice but by an even more dangerous, more powerful, virtually unstoppable force.

Bureaucracy.

A drama had played out over the course of the last half year, pitting In the Footsteps of Scott against a government policy determined to discourage private expeditions to Antarctica. Officials from the United States' National Science Foundation were particularly negative, raging from inhospitable to downright nasty.

The issue was our return from the South Pole once we had walked there. We requested space on one of the huge Hercules airlifters that customarily returned empty from supply runs to Amundsen-Scott Station. Surely there would be room in a cargo plane with a 45,000-pound payload for three half-starved trekkers? Of course, we realized if the American government helped us, it would set a precedent. They would be on the hook to help every private expedition. But in life I've learned that it never hurts to ask.

No, there would not be room. And no government cooperation whatsoever. So we moved on to plan B: Hire our own airplane. This was when Giles Kershaw, one of the great polar pilots of the modern age, enters the picture. He still today holds the record for the number of hours flown in Antarctica by a pilot.

Giles arranged to lease a Douglas DC-3 Tri-Turbo from Polar Research Laboratories in Santa Barbara, California. Problem solved. The Tri-Turbo Polair DC-3, nicknamed the *Spirit of Hope,* had been used extensively by the U.S. military before being sold to Polar Labs. We embarked for winter at the Jack Hayward Base with plans for our retrieval from the South Pole firmly in place.

Or so we thought. But that July, as Roger, Gareth, John Tolson, Dr. Mike, and I hunkered down in our hut on Cape Evans, the offer for the Tri-Turbo was mysteriously withdrawn.

The Polar Research Lab people gave a perfectly plausible explanation, but I saw the hand of the bureaucrat behind the decision. The U.S. Navy leased the Tri-Turbo for Arctic operations on lucrative terms. Did the "no cooperation" policy of the U.S. government extend to warning off Polar Research Lab? Perhaps by the threat of cancellation of the navy contract? A phone call was placed, and the Tri-Turbo denied to us.

Plan A, nixed. Plan B, thwarted. Plan C could be summed up by the old song title "God Bless the Child (That's Got His Own)." Giles Kershaw suggested a James Bond–style alternative: Dismantle an old single-engine Cessna 1-85, store it on the deck of the *Southern Quest,* off-load it, reassemble it, and fly from an ice floe.

This strategy sounded implausible, but it had the virtue of making the expedition totally self-sufficient, asking for no help from any government.

In September 1985, Giles laid out the plan to our business manager, Richard Down. To prevent further government interference, they kept the details secret, even from us, wintering over at Cape Evans.

The new strategy required a change in schedule that at the time seemed minor but would turn out to have fatal consequences. In order to be in Antarctica in January, in time to have Giles fly to Amundsen-Scott and pick us up, the *Southern Quest* would have to sail south a full month earlier than planned, in January rather than February.

The whole plan, the whole expedition, my whole life, came to smash in the pack ice off Beaufort Island. We had taken a risk and had very nearly pulled it off. The ship was going down only a quarter mile from the safety of open water.

Both of my Durham University Immortals, Pete Malcolm and Will Fenton, as well as Rebecca Ward were on the *Southern Quest* when it became clear she would sink. They had only thirty minutes to off-load onto the ice their personal effects, ship records, and vital supplies. The ship moaned and vibrated, iceberg impacting steel, *Titanic*-style. Pete made a quick inventory of what he would take and what he would have to leave behind. Winter clothing, yes. His treasured piano, no. Camera cases and movie gear, yes. Dishes and silverware, no. Emergency first aid, yes.

What about the $25,000 in cash that was in the *Southern Quest*'s safe? Pete thought about the stranded crew members on the desolate ice floe right outside the ship, referencing his knowledge of Shackleton's experience. *There'll be no use for money there,* he thought quickly, and left the cash behind.

An expedition member named John Elder blew a melancholy rendition of "Last Post" on his trumpet as the *Southern Quest* went down.

Promises

_____

What do you do when your ship sinks? Ship sinkings come in all forms, in business and in life. Stock-market meltdowns. Bankruptcy, whether corporate or personal. Health reversals, family issues, accidents.

The natural response in the midst of such disasters is to behave as a wounded animal does and go to ground. With elk, caribou, or wildebeest, for example, weakened animals are cut from the herd, left for the wolves or wild dogs to take. But we can do better than that. We're highly evolved animals, aren't we? We can triumph over our knee-jerk biological impulses.

The modern human equivalent of an animal going to ground is to forgo all risk. Protect yourself from further harm by adopting the safest of all possible strategies. But this is exactly wrong. Risk is an essential part of survival. It is one of the most difficult acts of faith to accomplish as a leader, but embracing an element of risk becomes a vital way out of extreme difficulty.

In January 1986, still quite stunned by my sudden reversal of circumstance, caught in the media glare like a deer in headlights, I neglected to grasp this essential lesson. I pulled back. Every time I ventured out of my thumb-sucking, fetal-position protection, I received a new blow.

I visited my banker. My sole asset with which to repay my personally guaranteed $1.2 million Barclays Bank loan lay off Beaufort Island, at the bottom of the Ross Sea.

"I am going to pay it all back," I said to Don Pratt.

"I know you will," he said.

Once again, I was lying. Or not telling the whole truth. I had not a thought in the world as to how a twenty-nine-year-old unemployed polar trekker with a degree in ancient history could climb out of a $1.2 million hole. I nodded solemnly to my banker, and he nodded dubiously back. My grandfather's core values of honor, duty, discipline, and honesty demanded that I resolve the debt. But the actual possibility of doing it seemed so remote as to border on the impossible.

I cursed my own idiocy and in my desperation cursed the decision of my Immortal friend Peter Malcolm in leaving $25,000 behind that was now at the bottom of the sea. That money could have been a sop to Don Pratt, it could have been seed money, it could have been *something*. Instead, I had nothing.

And what about my other debt, the promise I had quite grandly flung to the Beardmore winds? My "just don't kill us" plea? *I promise I will somehow do whatever I can do to protect you.*

Uh-huh. Right. Like all promises made in hostage situations, that one vanished like smoke. As soon as I was back in Old Blighty, I forgot about it.

First my banker, then my mother. I went to visit Em as soon as the media storm lulled a little.

"Good Lord, darling, what happened to you?" my mother said when first seeing me upon my return home. I thought she was commenting on my gaunt features. I had dropped twenty-five kilos—about four stone, or fifty-five pounds—on the South Polar Diet.

"Well, I just walked to the South Pole," I muttered. "You may have heard something about it."

"Yes, darling, but I mean your eyes. They've changed."

Under the glare of the Antarctic Plateau, my normally blue eyes had washed out to a pale gray. I hadn't noticed. I hadn't been around a lot of mirrors, and my two expedition mates weren't in the habit of noting details of my appearance. But leave it to a mother to catch the little things.

The slow learner still didn't put two and two together, linking a polar environment under assault by human-induced climate change to the rays that had fried the pigment out of my eyes. I thought it was simply one more Antarctica oddity. I recalled the American country singer Crystal Gayle, who had a hit song back then called "Don't It Make My Brown Eyes Blue."

But my eye color was more than a quirk. That year the massive hole in the ozone layer above the Antarctic became highly publicized. For seventy days we had marched beneath a sky stripped of its natural protective ozone, molecules of which normally blocked harmful UV rays from bombarding the planet.

The release of CFCs, by-products of the industrial age, used primarily in solvents and refrigerants and now more commonly called haloalkanes, had caused the ozone depletion that turned my blue eyes gray. Roger and Gareth had brown eyes and weren't affected.

In the mid-1980s, the hole in the ozone layer was big news, the inconvenient truth of the day. I wish I could say the scales fell from my eyes immediately, and right then and there I fulfilled my Beardmore promise, suddenly transformed into a champion crusader for the protection of the planet in general and the preservation of the polar regions in particular. That was not the case. Instead, like the *Southern Quest* getting swallowed by the pitch-dark, colossal-squid-haunted waters of the Ross Sea, I sank into the blackest of all black depressions. I drank. I contemplated my bankrupt financial status.

The Japanese polar explorer Nobu Shirase died penniless and forgotten. ("Study the treasures of the Antarctic," he exhorted, "even after I am dead.") Roald Amundsen, although widely celebrated, left massive debts behind at his death. Scott and Shackleton fared little better. Shackleton died broke, on a pointless Antarctic circumnavigation cruise in his ship *Quest,* a trip conceived primarily as a way to raise funds and pay off loans from earlier expeditions. Scott perished with a plea on his lips for others to take care of his wife and child, since they were bereft of any financial legacy from him.

Maybe I should have been honored to join such a long line of Antarctic bankrupts, but on the contrary I felt miserable and shamed.

Once again, it was Captain Scott who led me to take the first step out of my difficulties (just as, to be fair, he had led me into them). Not the symbolic British hero, either, but the flesh-and-blood, living, breathing, back-from-the-dead corporeal presence of the man himself.

Left Behind

Weighing on my mind equally with the $1.2 million Barclays debt—well, perhaps not equally, since nothing quite weighs on the mind like $1.2 million—was yet another promise I had made, equally rash in retrospect. I had solemnly sworn to my mentors, Lord Shackleton, Sir Vivian Fuchs, Jacques Cousteau, and—most of all—Sir Peter Scott, that I would leave Antarctica as I had found it.

"There's one condition for my support," Peter Scott had told me when I first came to him at his Gloucestershire home. "I want you to take your hut, your expedition camp, and anything else you've erected down there—I want you to take it all away."

I assured him I would. My promise must have struck him as facile, because he came back at me. "I'm serious, Swan," he said. "Nothing but beach. Footsteps only in the snow. I'm going to hold you to it."

Again, I said I would. On my grandfather's honor.

And I had made a determined effort to fulfill that promise. The *Southern Quest* would be on hand to take away every scrap of human crap that In the Footsteps of Scott scattered around at Cape Evans. Hut, radio tower, oil drums, everything.

But then suddenly the *Southern Quest* wasn't there. The spoor of the expedition still lay strewn across the beach, but I had no way to fulfill my promise to clean it up. My credibility was on the line.

In the midst of those fast, furious hours after we arrived at the South Pole, when the Americans were giving us the bum's rush off the continent, I witnessed one of the most selfless decisions I have ever seen anyone make.

Stopping briefly at McMurdo before being shuttled to Christchurch and then on to Heathrow, I felt the weight of the promise I had made to Sir Peter Scott. What would happen to our equipment at Jack Hayward Base? "I need someone to stay in the hut over the winter," I blurted out to the assembled crew of In the Footsteps of Scott.

Gareth Wood raised his hand. He didn't hesitate, either, but put his arm up right away. Here was a bloke who had just spent a punishing seventy days walking to the South Pole, whose feet had been frozen and blistered to the point of crippling him. Before that, he had been sequestered over the endless Antarctic winter in 16×24-foot quarters—the same hut in which he was now volunteering to spend another nine-month winter! Another year of his life.

After the pole, after the sinking, I was one stunned boy, but relief over Gareth's decision floored me anew. He wound up staying over with two other crew members, Steve Broni and Tim Lovejoy. Greater fortitude hath no man, than he who winters two years in a row in Antarctica.

Overwhelmed with emotion, I shook Gareth's hands. "Whatever I do, I promise I will come back and get you out of here, with all our equipment and rubbish, too."

My call for volunteers to winter over was in part a way to force me to keep my promise to clean up the beach at Cape Evans. I knew myself well enough to know that if I left the hut down there and departed for sunnier climes, my impulse would be to rationalize. I was already rehearsing my excuses to Sir Peter Scott and the others. Listen, I know I promised to clean up after myself, but my bloody ship sank. Circumstances beyond my control. Sorry about that.

And I think Peter Scott, Jacques Cousteau, Vivian Fuchs, John Hunt, and Edmund Hillary would have understood. The unbelievable sinking of the *Southern Quest* would have moved them. In public, to my face, they might have given my previous promise a pass. In private, though, I know they would have been disappointed.

We had done so well on the trek itself. Apart from human waste, we had hauled every bit of our rubbish out with us. There were days when this bag of garbage—margarine tins, fuel cans, packaging—enraged me. I wanted to strip every ounce from the incredible dead weight of my sledge. But Roger insisted. Today, the bag of In the Footsteps of Scott rubbish is in a museum in New Zealand, an object lesson on the proper way to run an expedition.

In the aftermath of his pioneering trek to the summit of Mount

Everest, Sir Edmund Hillary became terribly dismayed by the mess that littered the base camps. People lined up on the route to the top of the world's highest mountain as though it were a subway entrance. Discarded trash, oxygen bottles, and human waste desecrated the once-pristine Everest landscape. Climbers actually had to step over dead bodies on the way to the summit—the frozen corpses of those who had died on the mountain before them.

The detritus of Jack Hayward Base represented the first bit of trash from private expeditions marring the continent. I knew there was a possibility I might return to the United Kingdom and somehow go on with my life, making my peace with the fact that I had besmirched the pristine. But if I left people at the base, I knew I would have to return to clean it up. Asking for volunteers was my way of keeping myself honest, making sure I would do what I said I would do, which was scrub the beach at Cape Evans of our presence there. Jack Hayward Base was a bloody long way from anywhere. I might not return for discarded matériel, but I would have to for marooned human beings.

It was like getting that fifty-pence piece from the kid at the Docklands school. I was stuck with it. I had to deliver. Yes, I could have pocketed his coin and then blown him off, just as I could have left the hut at Cape Evans and wriggled out of my vow to my mentors. But something stopped me, and the values inculcated in me by my parents and grandparents were an element of it.

So here was another promise, this one flowing out of my original one to Sir Peter Scott. In order to keep the promise to remove all traces of In the Footsteps of Scott from Antarctica, I was forced to ask for volunteers to winter over. That led directly to my promise to Gareth, to find a way to retrieve him when his heroic second winter stretch at the hut was over.

The bad guys, unquestionably, were the only-by-the-rulebook bureaucrats of the National Science Foundation. Not the scientists and staffers at McMurdo, who were wonderfully warm and friendly to every person involved in our expedition. But Peter Wilkniss and Edward Todd of the NSF cannot be described as anything but totally narrow-minded.

America's McMurdo Station is actually classified as a ship, with a naval hierarchy in place. Captain Shrite was the military commander of McMurdo when the *Southern Quest* sank. Incredibly, the NSF bureaucracy instructed Shrite to deny our captain, Graham Phippen, even the customary courtesy of sea and weather reports.

We were also denied routine communication help, which meant not being allowed to talk to expedition members over the base radio when we arrived at the pole. On the flight in the Hercules to Willies Field at McMurdo, crew members were instructed not to speak to us. The prohibition was so bizarre as to be laughable, a churlish, Orwellian inversion of the normally collegial relations among Antarcticans.

Captain Shrite—we had taken to calling him Captain Shite behind his back—responded with brusque unconcern when we asked what we should do with our Cessna airplane, left at Cape Evans in the wake of the *Southern Quest*'s sinking. There was no way to fly such a small plane across the wastes of the Southern Ocean. The original plan had been to dismantle the plane and bring it back home the same way it had come, strapped down on the deck of the ship.

"Push it over the edge into the sea," Shite said.

I almost blew up at Shite right then and there. His attitude symbolized the casual approach to garbage and waste that I had seen at McMurdo and Amundsen-Scott Station. I remember how disgusted I was, after arriving at the pole, to see the personnel at Amundsen-Scott eating with plastic knives and forks. Afterward, a cook's helper would take the plastic implements and the food scraps outside and toss the whole mess into a hole in the snow. *It's a big place,* Shite's attitude seemed to say. We can afford to trash it. Dump it into a hole. Toss it into the sea. No big deal.

When the worst happens, when the shite hits the fan and your ship sinks, what remains important is your course of action as a leader. What I took away from the experience at Cape Evans was simple. *Control what you can.* I could not control the situation with the *Southern Quest*. I could not keep my ship from sinking. I could not control the NSF bureaucrats. They were going to restrict our options in whatever way they wished.

But I could control some things. The danger in extreme situations is getting overwhelmed by disaster. You throw up your hands. You forget there are always elements in any situation that you can manage. Find out what they are and put your efforts into those.

At Cape Evans, I could control what happened to the Cessna and to the equipment at our base. So that's what I did. I drew a line in the snow. NSF, the *Southern Quest,* no control over choices. Jack Hayward Base, control. We could only control our own credibility. We had to demonstrate that private expeditions could look after themselves.

We chose not to follow Shite's advice, leaving the Cessna, along with the hut, in the care of Gareth and his two mates.

The official ostracism of In the Footsteps of Scott by the National Science Foundation continued. The surly attitude of the McMurdo bureaucrats applied to Gareth and the others spending the winter of 1986 at Jack Hayward Base. They were made to sign a waiver agreeing not to expect any supplies from nearby McMurdo Station—no accommodation, no laundry facilities, no hospitality at all. Gareth, Steve, and Tim were prevented from being invited for a meal or even a beer at McMurdo. But they kept to themselves and made the best of a bad situation.

Gareth almost didn't make it home. A short expedition to the north, to the Shackleton hut at Cape Royds, almost ended in tragedy.

Gareth, Steve, and Tim cut across the sea ice of Backdoor Bay. A crack had refrozen into clear blue ice. Gareth stuck out a boot, tapping the surface to test its solidity. Suddenly a leopard crashed upward, shattering the ice, lunging out of the water to seize Gareth around the leg.

Leopard seals—Scott called them sea leopards—are among the most magnificent mammals of the Antarctic. A keystone rather than an apex predator, they are sometimes preyed upon by orcas, or killer whales, but nonetheless are the monsters of the South Polar seas. When Steve "the Crocodile Hunter" Irwin journeyed to Antarctica, as a connoisseur of predatory animals he naturally gravitated toward the leopard seal as his local favorite.

At ten feet long and one thousand pounds, with its array of inch-long canine-shaped teeth, the leopard seal clamped on Gareth's leg inspired no admiration, only terror.

"He must have been tracking us across the sea ice for some time," Gareth said. It is a disturbing image, a sinuous, snakelike seal looking upward at human shadows moving across the surface ice. The beast dragged a screaming Gareth toward the open water. Once submerged, Gareth would be dead. Steve tried to beat the seal back, attacking its head with the crampons on his boots. Blood gushed from human and seal both. Finally the animal loosed its grip, sliding backward into the sea.

Steve and Tim rescued their wounded mate and hauled him away from the hole in the ice. Suddenly the hole exploded in a huge spray of seawater and blood, and the seal again attacked. This time it charged bellowing across the ice, grabbing Gareth by the boot. Again Steve and Tim managed to fight it off. They pulled Gareth to safety.

When Scott's *Discovery* expedition shot and dissected a sea leopard, Wilson discovered twenty-four half-digested penguins in its stomach. The maculate brilliance of the beast, as dead-eyed and remorseless as a shark, to me mirrors the polar environment itself—inhuman but somehow bracing in its cold and alien inhumanity.

Johnny Mills

Weakened, gaunt, changed in eye and mind, weighed down by one rash promise after another, groping for answers in the aftermath of the expedition, I wandered bewildered around a borrowed office space, a place I wasn't supposed to sleep in but did, since I had nowhere else to stay.

London appeared to have changed utterly in the year-plus I was in Antarctica. After the contentless silences of the white continent, the clash and chaos of the city disoriented me. I felt as though I had climbed out of a tomb into a circus. History had galloped along in my absence. Our warehouse on the Thames, gone. The Canary Wharf project, the biggest development in Europe at the time, was transforming the Docklands neighborhood I knew so well into a thicket of cranes and construction.

I didn't know what to do. My tasks overwhelmed me. In hindsight, I believe I was suffering from post-traumatic stress disorder. At first I did what every red-blooded Englishman does when faced with catastrophe. I got drunk. That only added a series of brutal hangovers to my myriad problems.

"The noose of laurels" is the phrase that the British polar explorer Wally Herbert used to describe my predicament. You charge, charge, charge toward a goal. You attain it. There are complications. It's not like you imagined. To your dismay, you find that you remain the same old disappointing you. Now what?

I knew I could not put off a visit to my accountant forever. As I sat in the office of Ian Coombes, the news was so dark as to be almost laughable. I had dug such a deep hole that I could not envision how I would ever get out.

"We have a problem," Ian Coombes told me drily. "We have no money. Would you like to see a list of your debts?"

That was the last thing I wanted to see, but I nodded miserably.

Ian lifted up a hefty stack of computer printout paper, the sprocketed kind designed to feed through a printer continually. Then he climbed up on a chair. He unrolled the paper dramatically, and it spilled downward and rolled across the floor. The pages were closely printed with sums owed to my creditors.

I had already managed to cross two debts off the list, at least, paying back Pete Malcolm and Mark Fox-Andrews for their investments in the doomed *Southern Quest*. That left a mere $1.2 million, Ian informed me. Among other things, we had slipped out of the United Kingdom at the start of the expedition with a $250,000 tax bill chasing after us.

"Just let's try to keep me out of jail," I muttered.

Ian Coombes was always upbeat despite the grim news. The Institute of London insurance underwriters had refused to insure the *Southern Quest*. No one will insure a ship that far south. But incredibly enough, after the sinking the company came through with a donation to the expedition for half the value of the ship. It was enough to pay off Pete Malcolm's twenty thousand pounds and to repay Mark Fox-Andrews.

The sinking of the *Southern Quest* had generated enormous amounts of publicity. Something in the British soul reacts to failure more sympathetically than to success. It's a long-standing tradition. I was displayed like a king of fools in the public square, the story on every talk show and news program in Britain. The coverage was usually wry and tongue-in-cheek. Comparisons to the *Titanic* were made. Although they didn't come right out and use the word, "fool" was the message that the commentators conveyed.

Not everyone poked fun. All of us involved in Footsteps received a tremendous boost on March 29, 1986, when a letter to the editor appeared in *The Times* of London. "There have been a number of statements and criticisms of the 'In the Footsteps of Scott' expedition," the letter began, "both as to the competence of its members and their right to be in the Antarctic. We believe these criticisms are unfair and unjustified."

The missive went on, at length, to defend the whole enterprise, especially against the American government. But the real kicker came at the end. It was signed by the greatest living names of mountaineering and polar exploration: Lord Shackleton, Lord Hunt, Sir Vivian Fuchs, Peter Scott. And others, a who's who of the Royal Geographical Society. Letters with multiple signatures were not that common in *The Times*.

That letter of support represented a watershed moment. I realized that I had to lift myself up, dust myself off, and start all over again.

As the fundamentalist Christians ask themselves, "What would Jesus do?" I asked myself, "What did Captain Scott do?" After he returned from his first Antarctica expedition in 1904 on the *Discovery,* he was in much the same fix as I was, deep in debt and assailed on all sides by importunate responsibilities.

When the going gets tough, the tough go out on a lecture tour. That's what Scott did. I didn't have to read up on it—I already knew his life backward and forward. I realized the answer had been right there in front of me all along. Do what Scott had done.

But I also realized I was not a public speaker. I had seen my TV appearances. They were uniformly awful. Any train of thought I attempted got immediately derailed. I blinked beneath the harsh TV-studio lights. I stammered. Worst of all, my croak of a voice sounded like duck farts on a muggy day.

"I think you have to talk your way out of it," John Mills told me over the phone. On the basis of our kind correspondence begun at Jack Hayward Base, I had called him up. Our conversation quickly turned into a pathetic, one-sided catalog of woes recounted by yours truly.

"How can I talk my way out of it?" I said. "I'm not a public speaker. Johnny, what do I do?" We were suddenly on a first-name basis.

"Tell you what," Johnny Mills said. "Why not come up to my place? I'll give you a few presentation pointers."

So I journeyed to his beautiful red-brick mansion.

"Come to the bathroom, my boy," he said, after a cordial tea. "Look into the mirror."

I did so. I saw my own idiotic face, and, standing behind me, Captain Robert F. Scott.

You have to understand. To me, John Mills wasn't an actor. I had been schooled by repeated viewings of *Scott of the Antarctic.* This man *was* Captain Scott. Whenever I summoned my hero up in my thoughts, he had the face of John Mills.

And here my hero was, standing beside me, giving me tips on public speaking. Of all the surreal turns my life had taken—and there were a few of them back then—having the living image of Captain Scott give me advice ranked pretty far up there.

"What do you see in the mirror?" Johnny Mills as Captain Scott asked me.

"I see this person I don't like looking back at me," I said.

"Look at me," he said. "Do you think I like looking at this face? Do you think I like looking at me?"

I thought actually that he might like looking at himself. I thought most movie stars did. But I didn't tell him that.

"Speak into the mirror," he instructed. "What do you hear?"

"I hear a voice I don't like," I said.

"I hate my voice," he said. John Mills, Oscar winner, hating the sound of his own voice. "No matter how it sounds to others," he said, "to your own ears, your own voice never sounds very nice."

He told me to start from there. Every morning, I was to spend thirty seconds in front of the mirror, speaking to myself.

"I'll look like a complete fool," I said.

"Stop worrying how you look and start the process of becoming a good public speaker."

My own personal in-the-flesh Captain Scott gave me a lot of tips that day. Most essential: The first minute of your talk is the most important. If you don't hook the audience then, all is lost. Practice the opener of your speech over and over. Get it down cold.

I examined myself in Johnny Mills's mirror. The ghost of Con Scott stood behind me. "My name is Robert Swan, and I want to talk to you about what is truly possible as a human being."

"Again," the ghost of Captain Scott said. "Speak out, man! Say it like you mean it!"

"My name is Robert Swan . . ."

The voice of Johnny's wife, the playwright Mary Hayley Bell, came through the closed door of the bathroom.

"What are you boys doing in there?" Lady Mills asked.

"Working!" John Mills called back.

"Working," Lady Mills repeated. She didn't sound wholly convinced.

Leadership

It took me a long time, most of my early years, to get from that *Scott of the Antarctic* movie theater in Durham to the disaster at the dock. Now 2041 is the number by which I live, but the two-word phrase I live by is "sustainable leadership."

The question to which I keep returning again and again, on expedition and off, is how to maintain energy, initiative, and commitment over a stretch of years. From age eleven to twenty-nine, I somehow sustained the same vision: honoring Robert F. Scott by tracing his path walking to the South Pole. *I wonder if I could do that* morphed into *I am going to do that.*

I kept the dream alive back then without really knowing how I did it. In 2008, long after that first expedition, I journeyed up and down the west and east coasts of the United States in a sailboat called *2041*, advocating for preserving Antarctica, for climate-change issues, and for sane energy use. I stopped at numerous universities and colleges along the way, from Stanford and Berkeley to San Diego, talking to students.

I asked them all the same question: "How do you keep a dream alive?"

Again and again, students answered in remarkably similar ways. We are asked to care about so many things nowadays, they said. Social justice, racism, sexism, gay rights, antiwar movements, third world poverty, Darfur, consumerism run amok, disappearing rain forests. But as soon as we begin to care about one thing, to really embrace an issue, the students told me, suddenly that concern fades and another world-shaking problem crops up. Compassion fatigue sets in.

Their quandary was my quandary. How do you sustain leadership? I was asking them to care about yet another cause, the preservation of the last pristine wilderness on earth, Antarctica.

Perhaps, just maybe, I could engage a few of the students as long as we were face-to-face, by bringing into play every shred of power in my personality as well as exhibiting my dog-and-pony slide show. I could cast a polar spell.

But when *2041* pulled up anchor and I sailed away, what then? Energy fades, attention drifts.

What the students told me, again and again, sometimes using the same phrases, was it's all here today, gone tomorrow. Planned obsolescence has plagued not only products but ideas and causes. "Phone right now!" announces a TV pitchman. "You get this extra added bonus product." Another product. And another, and another.

But there is an alternate vision. Over the course of human history, certain people have accomplished incredible feats by focusing their energies over an extended period. That is the crucial secret that the history of polar exploration offers to reveal. How did Robert F. Scott sustain his leadership over a two-year span in the face of absolutely horrific conditions? How did he keep his team energized and on track?

If he failed or made a mistake, it wasn't a case of people simply losing face or money or position. People lost their lives. The same was true for Shackleton and Amundsen. The stakes were higher for them than for those in most enterprises, whether in business or life in general.

What's missing today in a lot of causes and movements is sustainable inspiration. First leaders need to make sure they are sustainable themselves, not burning out with effort, or ignoring the sustaining influences of friends and family, for example. We need sustainable leadership for sustainable inspiration.

When I present the 2041 mission to students and young people, they respond with incredulity mixed with excitement. Do you really mean we could be part of a three-decade-plus mission? It's the opposite of the here-today-gone-tomorrow efforts they have encountered all their young lives.

For seventeen years, from the time I first conceived the vague, schoolboy notion of following Scott on his trip to the pole until the time I made that first wharf-battering start to the expedition, I managed to sustain my focus. If you had asked me back then how I did it, I would not have been able to answer you. Multiple blunders and repeated hiccups of disaster showed that I did not know what I was doing. Somehow we prevailed.

As I've indicated, I'm a slow learner. But I've made it my life's purpose to uncover the secrets of sustainable leadership. I think of Scott as a man carrying a lit candle across the frozen landscape of Antarctica during a blizzard. His job was somehow to keep that candle lit. That was my job, too. How did I do it? Well, I didn't burst out of the house after seeing *Scott of the Antarctic* and start marching to the South Pole. The idea took a long time to germinate fully.

Broadly considered, my Scott expedition was a ridiculous undertaking—a twenty-something nobody raising five million dollars to embark on a useless quest. But the steps I accomplished to make it happen, when broken down afterward, actually seem quite simple.

Get people to help me.

Get money to support the effort.

Get it going and keep it going.

Do something larger with it.

That simplicity is deceptive. The path of true adventure never runs smooth. How did the expedition to the South Pole come together? Snafus, mishaps, and dead ends were matched by an equal portion of unbelievable luck, serendipitous connections, and happy coincidences.

That's what I would talk about when I hit the lecture trail. I would tell the only tale worth telling, the one about triumph against odds, the one that points the way to what is truly possible to achieve in this world. Joseph Campbell has identified it as the universal hero's story. It is Robert F. Scott's saga through and through, and, to a much lesser degree, the tale of In the Footsteps of Scott.

My story would be about disaster, but it would also be about leadership.

Giles

———

Istill had three men on the ice. The situation constantly nagged at me. Discarding one unworkable plan after another, I tried feverishly to figure out how to get Gareth and his mates out of there, and how to clean up Jack Hayward Base.

Pete Malcolm was starting fresh in Australia, working with Greenpeace. Like Zeus conveying a bolt from the blue, he forwarded a message from the international environmental group. Greenpeace wanted a presence in Antarctica to keep an eye on the Japanese whaling fleet, which was increasing its kill in open contravention of treaty accords. The organization also wanted to motivate the clean-up of the McMurdo bases, which was accomplished the next year. If we would allow Greenpeace to use Jack Hayward Base the next season, they would not only use their ship to get our people off the continent but they'd also scrub the beach at Cape Evans of all traces of our presence.

As far as I could see, the arrangement would mean everybody won, the whales not the least of all. Gareth would get out in December. The hut would be removed at the end of the season, in late 1987. Problem solved. A load was lifted from my shoulders.

Except . . . Murphy's Law was invented for Antarctica, where if anything can go wrong, it will. Later that year I got a message from Greenpeace. *Ice too thick. We can't get in. No approach to Cape Evans possible.*

Their ship, MV *Greenpeace,* was blocked by pack ice near the spot where the *Southern Quest* went down.

Back to square one. I consoled myself by thinking about Shackleton's incredible effort to get his men off Elephant Island after they had been shipwrecked. His astonishing open-boat journey to get help had become an iconic example of leadership. Holding tight to Shackleton's example of perseverance, but without a thought in my head as to how to proceed, I again sorted through my options.

It had worked before, so I again turned to Giles Kershaw. He
hatched an outrageous, epic scheme to fly across the whole continent in
order to extract our trio of maroons at Cape Evans. Dick Smith, the
Australian electronics magnate and aviation pioneer, signed on as a
sponsor of the effort. He had claimed fame for the first solo round-the-
world helicopter flight. Good on him, I'd say.

Of course, Gareth and the others could have just walked the twenty
miles from Jack Hayward Base to the big New Zealand and American
bases at McMurdo. They could have begged a ride home from there. It
would have been simple, but we refused to do it that way.

Still smarting from the Americans' boorish behavior when the
Southern Quest went down, neither I nor Giles wanted to have anything to do with them. We would accomplish the extraction and
retrieval of our personnel ourselves. As I've said before, never underestimate the motivating power of "Screw you!" I believe that's partly
what moved the iconoclastic Dick Smith to get involved, too. But all of
us fully believed we should uphold our honor, walking the talk to do
the right thing in Antarctica.

Giles's plan: Fly from Punta Arenas in Chile across the Drake
Passage to the Antarctic Peninsula, down the peninsula and across the
continent to the barrier, across the barrier to Cape Evans. The four-
thousand-mile route would bring us over Palmer Land, Ellsworth
Land, and Marie Byrd Land, past the continent's highest point, the
Vinson Massif (16,066 feet/4,897 meters high), and past its lowest, the
Bentley Subglacial Trench (8,333 feet/2,540 meters below sea level), in-
visible beneath the ice cap.

Four thousand miles in a Twin Otter airplane requires refueling.
We needed help, logistical support from pretty much anyone other than
the American government.

I decided to play the Shackleton card. I contacted General Lo
Peotegi of the Chilean Navy. Listen, I told him, your people supported
Ernest Shackleton when he went to retrieve his people from Elephant
Island. Don't you think it would be right for you to support us as we at-
tempt to retrieve our people from Cape Evans?

It was the same historical gambit I had employed to get Shell to donate six hundred tons of diesel fuel to In the Footsteps of Scott. The government of Chile gave us their support, in the person of Alejo Contreras, who would station himself midcontinent with 120 barrels of aviation fuel.

Alejo was a wonder. He knew nothing at all about Antarctica back then. But over the ensuing decades he would become a bearded, weather-beaten veteran of the mountains, one who knew positively everything there was to know about polar weather, polar transport, polar living, polar wildlife.

At that point I had never met Alejo, but Giles and I were trusting him with our lives. By the time we approached the Patriot Hills refueling station in Ellsworth Land, where Alejo waited, we would be running on vapors. If he wasn't there, we would be stuck.

From the very beginning, it was madness. Giles had filled the plane with countless bladders of aviation fuel. I climbed in and almost passed out from the fumes. A canary would never have survived.

"Ready?" Giles called out, giving me a grin that somehow dismissed the craziness of it all. I gulped and nodded.

Five hours later we were over Ellsworth Land. We had flown past the point of no return. The needle on the fuel gauge red-lined. Somewhere down below, beneath an impenetrable black cotton veil of angry clouds, a lone mad Chilean was supposed to be posted with our fuel.

"Where do you think we are?" I yelled to Giles over the clatter of the engines. "Somewhere near this fuel dump?"

Giles shook his head. He pointed to the Omega navigational instrument, with a satellite-based grid that was supposed to plot our course exactly. It indicated that we were somewhere over Bangkok. Satellites, compasses, and GPS receivers tend to go loopy at the bottom of the world.

"This is a bit hairy," I said. Giles nodded gleefully.

He dipped below the clouds, and there, glistening the dull gleam of the otherworldly Antarctic light, was the blue-ice runway of Patriot Hills. Latitude 80°18' S, longitude 81°21' W. Amazing. A small khaki-colored dot—a sole tent—showed near the runway.

Giles se. the Twin Otter down. I saw Alejo for the first time. He emerged from the tent looking as if he had been born from the ice and rock of Antarctica. Working manically at the hand pump, he refueled us. *Pump, pump, pump,* good-bye.

Flying over the emptiness of Marie Byrd Land and the Great Ice Barrier, I realized that if I had made this same flight before In the Footsteps of Scott, I would have never attempted the trek to the pole. The expanse below rolled out as though it was immune to the physics of distance. The magnitude of it sucked the soul right out of you. I knew I wouldn't have been able to summon up the heart to attack it by foot, if I had seen it first by air.

We landed on the ice right in front of Jack Hayward Base. It slowly dawned on me how important Giles's insistence on caching fuel there had been. The ten barrels of aviation fuel had made all the difference. We could not have made the rescue flight otherwise. How had he known?

When I asked, he shrugged. "In Antarctica, nothing ever sorts itself out as planned," Giles said. "You can count on it."

After the refuel at Cape Evans, and after packing Gareth, Steve, and Tim into the plane ("See? I didn't forget about you guys," I told them), we flew back along the same route to land at Patriot Hills again. There was Alejo once more. *Pump, pump, pump,* good-bye. After fourteen hours of flying, we landed at Punta Arenas. The lonely hero Alejo wasn't picked up by a Chilean military flight until three weeks later.

The whole effort was a Shackleton-haunted enterprise all the way. Giles and I did nothing to approach Shack's miraculous sixteen-day journey from Elephant Island to South Georgia in an open boat stuck together with canvas and dried seal blood. But his example helped us get through.

I did it to uphold the right of private citizens to go to Antarctica. I did it to thumb my nose at the Americans who had treated In the Footsteps of Scott so shabbily. But most of all, I did it because no one should go to Antarctica looking to be taken care of by other people.

I was greeted in Punta Arenas by a wire from Greenpeace. *The ice has gone out,* it read. *We can pick up your people after all.*

Icewalk

Fifteen minutes of fame is short. The media glare shone brightly upon In the Footsteps of Scott, illuminated us for a brief moment, then winked out.

But people remembered. The man in the street remembered. People came out by the dozens, by the hundreds, to hear me speak about our adventure. Francis Spufford, in his book *I May Be Some Time: Ice and the English Imagination*, explores my countrymen's complex national obsession with the Arctic and Antarctic. On my speaking tour of the United Kingdom—which eventually expanded to Australia and elsewhere—I saw firsthand that the public fascination was still there.

I spoke primarily about In the Footsteps of Scott. Gradually though, in response to questions from the audience ("You must have hiked underneath that hole in the ozone layer, didn't you?"), I began to talk more and more about the Antarctic environment.

In those days of the late 1980s, few people were seriously considering issues of climate change, but I was. I talked about how hydrocarbons emitted as a by-product of modern, industrial civilization had ravaged the poles. How chemical toxins inevitably migrated to the ends of the earth. How a precipitous rise in global temperature would put New York and London under water.

But I didn't really believe any of it. I spoke about it merely because the environmental message seemed to go over well. If my audience had been obsessed with penguins, say, I would have featured them in my talks.

I knew I couldn't very well stand up in front of people and say, "My name is Robert Swan, and I'd like you to help me pay off my debt." That would have been closer to the truth of it. And that's really what I was doing out on the lecture trail.

Looking back, I am somewhat ashamed. I was young and desperate. But this brand of bad faith—we call it "greenwashing" now—is the bane of the movement. I wasn't thinking about Antarctica. I was thinking of myself.

Over the course of 1986–87, I traveled fifty thousand miles around the United Kingdom giving lectures. But chipping away at a mountain of debt with speaking engagements just wasn't doing it.

At UNESCO headquarters in Paris, I met up with Jacques Cousteau. "We've been at this a long time, Sir Peter Scott and I and a few others," he said. "It troubles you when you think no one's listening."

What I had to do, Cousteau told me, was develop short-term missions that the public could get behind. If every one of his ten dozen fabulous TV documentaries had been general—Save the Ocean!—then the viewership would have been be nil. So he did programs on the shark, on the dolphin, on the wonders of the sea. The long-term goal, that of protecting the ocean, was concealed inside Cousteau's entertaining, short-term personal adventures aboard the *Calypso*.

Peter Scott told me much the same thing using different terms, when I visited him at his Slimbridge home in the United Kingdom. Divide the overall objective of preserving Antarctica, he said, into manageable projects that you can handle over the short run. "You'll be surprised, Robert, how quickly the short term resolves itself into the long term," he said, speaking to me as a man of seventy-five years. "It seems to happen almost overnight."

What I needed to find was an immediate goal I could focus my energies upon, one that would help illuminate the larger, more expansive goal of fending off threats to the polar environment. I began to have a terrible thought—that the only way out of my dire circumstances might be the way I had gotten in. I tried to push it out of my mind, but there it was.

Another expedition.

Casting about for what I was good at, I realized it consisted pretty much of one thing: firing people up for large-scale projects, presumably the more absurd, the better. What else could I do? What else was I good for?

The idea for another project began to prey on my mind. I would walk to the North Pole just as I had done to the South.

Not in the same way, not a mad dash with a three-man team. Since the problems plaguing the polar environment were global, I wanted this

North Pole expedition to be global, too. I wanted to attract people from countries around the world. (I also knew this would increase debt-paying opportunities. I admit to being just that craven.)

The idea had actually occurred to me, and been thrust away from my mind, during the early stages of In the Footsteps of Scott. No one had ever walked to both Poles. It was one of those finely parsed, thinly distinguished polar "records" waiting to be established. *If I don't do it,* I thought at the time, *someone else will.* Then I buried the thought and got on with the expedition at hand.

But now it rose to the surface again. Most of the dedicated souls I'd been with on Footsteps of Scott ran screaming from me as soon as I broached the idea.

Rebecca was the first to step back, not only from the expedition but from my life. She had dedicated five years of her life to Footsteps. "One pole is enough for me," she said. "I can't deal with you doing two."

Pete Malcolm bowed out right away. "I just can't do another one," he said. I think a little part of Pete had gone down with the *Southern Quest* (along with the expedition's cash). He'd loved that ship.

Pete's Greenpeace mates, our saviors in the good ship MV *Greenpeace,* had packed up and carted off our whole kit from Jack Hayward Base: the hut, generators, radio antenna, and trash. The Greenpeacers had, in their words, "polished the stones" at Cape Evans, removing any trace of Footsteps of Scott.

Two of my rash promises had been fulfilled: my pledge to Gareth to come back for him, and the commitment I'd made to Sir Peter Scott, Jacques Cousteau, and Lord Shackleton to leave Antarctica as I'd found it.

Now there was only my obligation to Don Pratt of Barclays on the Strand. A little matter of a million-plus. And, oh yes, there was that other promise, that vague pompous oath sworn to Antarctica itself.

A promise to a continent? How was that supposed to work? If I didn't keep it, was Antarctica going to hunt me down and exact its revenge? Not likely. Was going back to collect the hut enough? Maybe it was. But deep down I knew it wasn't. I tried to put the whole matter out of my mind.

Work was already in progress toward inspiring the preservation of Antarctica. With many of the Immortals not taking my calls, it was time to move forward. The publicity surrounding In the Footsteps of Scott, ship sinking and all, helped immensely.

One of the first persons I contacted was also one of the best, a former Royal Marine named Rupert Summerson, a topflight organizer and navigator. I'd met Rupert where I met most of my victims, on the 1981 BAS trip to Rothera. I had to be reminded that he had once dated my sister Lucinda in Britain, way back when she was fourteen years old.

While I traveled the world raising money, Rupert toiled in the nitty-gritty business of planning. Richard Down came on board again as the organizer of fund-raising. Suddenly we had a name for our effort—Icewalk. And one of those convenient handles so beloved by modern-day adventurers. I would be "the first person to walk to both the North and South poles."

I didn't really *want* to be that person. I hated being cold. I hated walking. "When I get home," said Titus Oates, a member of Scott's South Pole expedition who in the end never did get home, "I'm never going to walk another step."

Now I was going to walk another five hundred miles? But I could see no other way out of my predicament.

My world became a blur of travel and supplication. There was no money in the pot yet. During this time I borrowed from family and took some from my lectures fee. I put that into travel. I stayed with long-suffering friends in different cities and towns.

I remember endless cheap flights, sleeping the night at bus stations, changing clothes in public restrooms, carrying my suit wrapped up in plastic so it would be fresh the next day. After meetings, I would get back into my casual travel outfit of shorts and T-shirt and carry on. It was hand to mouth. I had a girl in every port. I had to depend on the kindness of strangers. Raising money for the effort was one of Australia's top businesswomen, Christine Gee, aided by Liz Courtney and Christine McCabe. We were supported by Mark Fish and inspired by Australia's bestselling author Bryce Courtney.

A breakthrough came when an eccentric Brit in Japan named An-thony Willoughby, who shared a speakers bureau with me, read an ac-count of Footsteps of Scott and invited me to meet with him. I arrived in Japan, totally jet-lagged, and together we began to do serious dam-age to a bottle of whiskey.

"By the way, dear chap," Willoughby said, "the Amway people will be here shortly."

I looked at him, bleary-eyed, barely holding on. The who?

Willoughby wanted to introduce me to the president of Amway in Japan. The Japanese arm of the direct-sales household-products com-pany was one of the first companies to recognize the value of green products. It was fabulously successful, but it had a very real social im-pact, too. An Amway career was among the only ways that Japanese women could make a living on their own, out from under the oppres-sive thumb of male-oriented society. Amway represented one of the few paths to female independence.

I did not comprehend all this that first night. In fact, I was pretty far from comprehending anything apart from the rotary velocity of the room. Hadn't we better postpone a business meeting until I was less in my cups?

But I underestimated the traditional way of doing business in Japan. The head of Amway Japan walked into the room with his assis-tants, took one look at me and Willoughby, and said, "Well, I see that we have some catching up to do."

He ordered another bottle and we continued our "business meet-ing" in the alcohol-drenched fashion of the Land of the Rising Sun.

Rupert Summerson's quiet, methodical approach to the expedition complemented my wilder, more emotional style. He was especially con-cerned with how we picked our team members. Rupert believed we must choose our polar companions carefully, because, he said, in a case of extreme duress we may wind up having to eat them.

Beginnings

My father, Douglas Swan.

My mother, Margaret "Em" Swan.

At age eleven: the only time in my childhood I ever got caught wearing a suit.

Heroes

Captain Robert Falcon Scott
(June 6, 1868–March 29, 1912), Royal Navy.

Ernest Shackleton (February 15,
1874–January 5, 1922), Scott's
subordinate and sometime rival.

The brilliant Norwegian
explorer Roald Amundsen
(b. July 16, 1872, disappeared June
1928), first to the pole by four weeks.

In the Footsteps of Scott Expedition

Glory days: Pete Malcolm *(right)* and me aboard *Southern Quest* in the summer of 1984.

My ship came in with a vengeance: we crashed into both the pier and the century-old Tower Bridge in London.

Jack Hayward Base at Cape Evans, our home away from home for nine months.

The interior of our hut: *(left to right)* Dr. Mike Stroud, Roger Mear, me, Gareth Wood, and Captain John Tolson.

Roger and me sledging early on in the trek—picture taken by Gareth.

In the footsteps of Scott.

JAN 11TH 1986
23.48hrs

End of the line, January 11, 1986: the *Southern Quest* sank on the same day we reached the Pole.

Bellingshausen Cleanup

The beach at Bellingshausen before the cleanup.

The beach at Bellingshausen afterward.

The Overland Journey

The sailboat *2041* being trucked through South Africa.

Voyage for Cleaner Energy

The Voyage for Cleaner Energy kicks off as *2041* sails near San Francisco's Golden Gate Bridge.

Inspire Antarctic Expeditions

An IAE group in Zodiac.

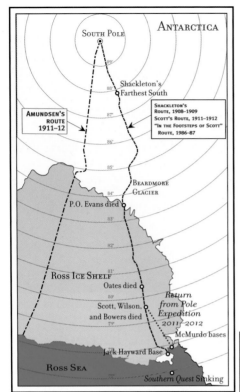

In the Footsteps of Scott: We trekked 873 miles in seventy days, man-hauling supplies and without radio, across Antarctica to the South Pole from November 3, 1985, to January 11, 1986—the longest unassisted march in history.

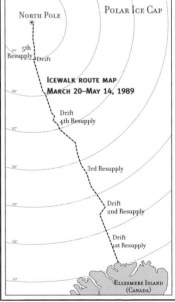

Icewalk: More than 600 statute miles across the ice cap were traversed to reach the North Pole on May 14, 1989, with eight trekkers from seven nations.

Amundsen

Leadership is about selection. You have to assemble the right people around you. Those right people may or may not be your best friends. In some cases, they can be those who rub you exactly the wrong way.

My simple rule for survival and success on a polar expedition: Take along an Amundsen.

Norway's Roald Amundsen must be accorded the mantle of greatest polar explorer of all time. He achieved a pair of firsts—first through the Northwest Passage, first to the South Pole—that had long bedeviled humankind. His approach—tireless preparation, imitating the best methods of indigenous peoples, dependence on strict logic stripped of all emotionalism—set the bar for all expeditions to follow.

Amundsen stated his precise prescription for success:

> I may say that this is the greatest factor—the way in which the expedition is equipped—the way in which every difficulty is foreseen, and precautions taken for meeting or avoiding it. Victory awaits him who has everything in order—luck, people call it. Defeat is certain for him who has neglected to take the necessary precautions in time; this is called bad luck.

In almost every field, some humans seem to excel above all others of their species to such a degree as to be almost unworldly. Picasso in painting, Mozart in music, Nelson in naval tactics, Amundsen in polar exploration.

Here's a fact about Amundsen. In 1909, sitting in his study in Norway, planning his assault on the South Pole, he wrote this sentence in his notebook: "Thus we will be back from the Polar journey on January 25." Three years, thousands of miles and millions of variables later,

Roald Amundsen and his companions were indeed back from the pole on January 25, 1912. Incredible.

In the Footsteps of Scott had its own Amundsen in the person of Roger Mear. Roger presses every button on me. He can yank my chain without even trying. Sometimes he makes me think I'm *all* chain. But I would die for him, and he was the person who got us to the South Pole.

Icewalk would have its own Amundsen, also, in Dr. Mikhail "Misha" Malakhov, one of the top Soviet polar explorers of the day. Misha was just as stern a taskmaster as Roger. They shared the maddening characteristic of being always right. They were both polar geniuses and numbered among my own personal Immortals.

Misha would soon become one of the first official Heroes of the new Russia. It was the era of glasnost and perestroika, a new openness in Soviet society, and the internationalism of the Icewalk expedition appealed to Misha. I was determined that global challenges required global approaches.

In the spirit of that, our Russian Amundsen was matched by a West German explorer named Arved Fuchs; a Melbourne schoolteacher (and world-champion canoeist) named Graeme Joy; Japanese mountaineer Hiroshi Onishi (who would die in an avalanche in 2007); and a Canadian Inuit named Angus "Gus" Kannerk Cockney.

On a trip to the States to speak at Manhattan's Waldorf-Astoria to the city's famed Explorers Club, I challenged that august institution to widen its scope. From the dais, I looked out at the room. Not a black face anywhere. I probably would not have remarked on that fact were I not in the middle of reading a book about Matthew E. Henson, one of the world's most forgotten explorers, yet a man who probably was the first to reach the North Pole.

The Explorers Club presented me with a flag to bring on Icewalk, but I came back at them with attitude.

"Your Explorers Club flag has flown in almost every square inch of inhospitable terrain on the planet," I told the group. "Ladies and gentlemen, it is a high honor to be asked to carry it to the North Pole. However, no one has mentioned the great black American explorer Matthew E. Henson. To your credit, you made him a member of the

Explorers Club. Your flag has been everywhere. But it hasn't reached into some of the world's toughest regions, just a few miles from here—Harlem and Brooklyn and the Bronx."

I thought that the Explorers Club must hate me, that I had embarrassed the entire group by pointing out the truth. This was the period of *Fort Apache, the Bronx,* an iconic movie that defined a certain kind of urban wasteland.

I was surprised when the next day the Explorers Club president, John Levinson, invited me along on a driving trip through the South Bronx, past blasted-out buildings, rubble-strewn lots, poverty, and hopelessness written on the faces of passersby. *Wouldn't it be good,* I thought, *to bring someone from this environment on Icewalk?*

"We've arranged for you to talk to two uptown schools," Levinson told me. "We'll also help you find and train a person from the minority community in New York City for your expedition."

I thought about Henson, an African American civil engineer who was with Commander Robert E. Peary's 1909 Arctic expedition, which claimed to be the first at the North Pole. Henson actually gained the goal first. Peary, crippled by frostbite and at that point riding in a sled, came soon after but was celebrated by the world as first man to reach 90 degrees north.

Peary and Henson had met during pioneering work on the Panama Canal. The navy man was impressed by the younger sailor's mastery of navigation. They were never apart after that.

Henson's later efforts to claim priority angered Peary, an inveterate glory hound who considered the experienced Henson no more than a servant ("my dusky companion," Peary called him). Henson disappeared into history, spending his remaining years as a customs clerk in New York City. Peary's claim, controversial as it was, achieved wide recognition. To my mind, he and Henson got bloody near enough.

An heir to Henson's legacy presented himself in the person of Daryl E. Roberts, twenty-three years old. A Brooklyn native, Daryl saw the ravages of crack cocaine in his community and wanted to accomplish something to provide a positive role model for children. He had no polar experience and little outdoor know-how at all. But the possibility of honoring Henson by including an African-American on the Icewalk expedition, along with Daryl's natural bonhomie, persuaded me to enlist him.

"At least I won't get sunburned," Daryl joked. He would become the first American and the youngest person ever to have walked unassisted (without dogs) to the North Pole.

So, eight Icewalkers from seven countries: myself and Rupert Summerson from the United Kingdom, Misha from Russia, Graeme from Australia, Hiro from Japan, Arved from West Germany, Gus from Canada, Daryl from the United States. An international group indeed.

Amway Japan came on board as the expedition's major sponsor, along with the world's bestselling newspaper, Japan's *Yomiuri Shimbun*, with a circulation of sixteen million per day, and its youth edition, which came out twice weekly and also had a circulation of sixteen million.

The path to secure *Yomiuri*'s sponsorship was interesting. At our first meeting, I was in a room with eighteen people, mostly midlevel executives, editors, and public relations experts. Six months later, after a series of meetings that grew increasingly smaller as I climbed the ladder of corporate authority, I sat alone with the publisher himself, just the two of us. We sealed the deal.

To accentuate the global theme—and to increase the logistical complexities of the expedition exponentially—I created another component of Icewalk. We devised a plan to bring an international group of students north and gather them at a research base in the Arctic while Icewalk was in progress. The students would work on environmental projects and monitor our advance across the ice to the pole.

Following in the direction of Jacques Cousteau and Sir Peter Scott to engage and thus inspire young people, we assembled twenty-two students from fifteen countries at Eureka, Environment Canada's base on Ellesmere Island in the high Arctic.

The impulse for all this grew out of my talks on In the Footsteps of Scott. When I spoke to young audiences, I got two almost contradictory responses. A determination to address environmental problems that far outstripped anything I experienced in adult audiences, and at the same time a feeling of remoteness, passivity, and paralysis. I knew that if I could just get young people to come into contact with the polar environment firsthand, they would become highly effective ambassadors for change. My vision was for a cadre of environmental blade runners, spreading the message when they returned to their home countries.

Besides, it all made for good copy. I don't want to overstate my cynicism during this period. Of course I believed in the preservation of the polar wilderness. Of course I sought to protect it against the assault of industrial toxins and pollution. But there is a difference between a faith one espouses and a faith that has taken root in one's soul. Even as I organized such idealistic projects as the Eureka student gathering, in truth I had a conflicted heart.

I set out with my seven companions on March 20, 1989, from our advance base at Cape Columbia on Ellesmere Island. Since we would be resupplied en route, our kits were lighter than those for Footsteps but still a healthy 150 pounds. I man-hauled a long sledge similar to the one I used in Antarctica, and so did Arvid Fuchs, while others chose smaller sledges or backpacks. We skied the short slope of the shoreline onto the ice of the Arctic Ocean, which would be our home for the next two months. We started off five hundred miles from our goal, the North Pole.

Teddy, the little bear that had journeyed with me on Footsteps of Scott, was once again tucked inside my jacket, a talisman of good-luck gingham. If we made it, he would be the first bear to journey to both poles.

I also carried with me a memento from our resident psychic, Emma Drake—a small piece of wood from the hull of Scott's original ship, *Discovery*. She had pestered workmen who were restoring the vessel as a visitors' attraction, gotten them to give her a sliver, and placed it in my kit with her good wishes. I had no idea how crucial and prescient her gift would prove.

As soon as we hit the ice, we ran into trouble. The Arctic is an active ocean. The constant movement of the sea ice jams massive, shattered slabs of it against the shoreline. A brutalized landscape stretched ahead of us, as if there was a hurricane going on and God suddenly froze it into icy rubble strewn haphazardly across our path. It wasn't even a path. It was an obstacle course.

I found myself actually missing the pan-flat expanses of the Great Ice Barrier and the Antarctic Plateau. Regret over embarking on the trek obliterated my resolve. I was lucky the others could not see my face, wrapped in its protective hood. My expression was one of sick despair.

Walking to the South Pole, I had not really known what I was getting myself into. The task remained abstract, no matter how much I read Scott's journals and stared out at the polar landscape around the hut at Jack Hayward Base.

But during those first few thrusts onto the ice off Cape Columbia, the awful reality of what lay ahead overwhelmed me. I knew all too well the agony of a trek such as this. It was like retaking a horrible exam. My mind turned mulish. I simply could not face the days that lay ahead, the pain and effort, but most of all the soul-numbing monotony.

I felt like collapsing, or screaming. Instead, I bent over my ski bindings and wept, totally overcome. My tears dropped onto the ice of the Arctic Ocean and froze there.

Eighty-fourth Parallel

Tears didn't matter. They turned to ice, my breath froze, my snot froze, my sweat froze inside my clothes, turning them into a particularly uncomfortable kind of ice armor. My pee froze almost before it hit the ground.

I was reminded of the reports from the Scott expedition, of those last despairing marches, when Taff Evans and the others simply shat inside their pants as they trudged onward, letting the turds freeze and then shaking them out of their clothes afterward.

An unpretty picture, but the first days of Icewalk rivaled it for pure ugliness. The frozen ocean ran to the horizon. Far out from the shoreline, the tidal surge pushed and crumpled blocks of ice into fantastic shapes, the polar opposite, if you'll forgive me, of flat.

Beneath our feet was the black sea—and not a thousand feet below, as it was on the Great Ice Barrier in the Antarctic, but only a half dozen feet, sometimes less. The blocks and slabs of ice moved with the ocean's flow, groaning and scraping as they rubbed up against one another. We were ants, crawling among the debris of a bombed-out city of ice. A city of ice that moved.

We struggled. Heaving ourselves up a tilted ice block, crashing down the other side, only to face another obstacle that was yards high. Here was a species of physical effort I had not imagined.

"Much harder," I wheezed out to Misha, "much harder . . . than the South."

The good Russian doctor looked at me strangely. He glanced behind us. We were still in view of the shoreline.

Daryl Roberts actually started out the trip with frostbitten toes, the result of a single reckless exposure to the elements during our stay in the Inuit village of Iqaluit.

I had learned from bitter experience that on walking expeditions, the effort of putting one foot in front of the other is only half the battle.

The other half is foot health. Every night after a trek came the process of drying, dressing, cleaning, attending to any hint of blister, frostbite, or wound. Our titular god was Dr. Scholl.

As the youngest and least experienced member of our group, Daryl suffered disproportionately. He developed an enormous blister on his right heel from walking without an appropriate insole in his ski boot. The blister festered and refused to heel. Walking was agony.

One evening, when we were one hundred miles from the pole, as Daryl removed his sock, the hard callus of his heel dropped off his foot, a ghastly lump that landed on the floor of the tent. Misha was in attendance, so Daryl could not see what had happened. "What was that?" he asked.

Misha murmured something in Russian. He picked up the pad of dead skin and hid it from Daryl, immediately binding the foot in heavy-weight gauze.

Misha was one of the world's top experts in frostbite, but before Icewalk he had never treated a black person. He had researched the subject before the trip and found the literature embarrassingly slight. He was forced to go back to the African American experience in the Korean War for relevant examples, and still the data was thin. Even though Misha dressed the wound each night (being careful not to let Daryl himself see how bad it was, lest he become depressed over it), the heel never healed for the duration of the walk.

Still Daryl persevered. We all did, though most of us were suffering in some way. An old back injury from my days on the Sedbergh rugby team came back to bedevil me. Arved came down with the flu and for a period during the march would have to stop every once in a while to vomit.

Hiroshi, the mountaineer, proved himself superb whenever any climbing was involved but became slow and exhausted with the strenuous, day-to-day exertion of sledging.

Daryl once tumbled into a snow-filled gulley. The oblivious, nose-to-the-ground Hiro skied right over him, unnoticing. "Thanks, Hiro!" Daryl called, raising a ski pole in salute from beneath the snow.

The saga of the great polar explorer Sir Wally Herbert occupied my thoughts. In 1969, Herbert traveled by dogsled across the whole of the

Arctic, from Point Barrow, Alaska, to Spitsbergen, Norway, one of the greatest journeys ever made and one that will probably never be repeated. But Herbert arrived on the Norwegian island on the same day that Neil Armstrong and Buzz Aldrin landed on the moon, so you've never heard of him. Throughout my early struggles on Icewalk, I meditated on Herbert as a symbol of the vagaries of fame. What was I doing it all for, if a real hero such as Herbert could be erased from the pages of history by a simple mishap of timing?

Psychological problems plagued us. The Inuit have a name for the depression that hits people in the polar regions, a phrase that translates into "the weight of life." Rupert became withdrawn. He had trouble with his metabolism and suffered continually from hunger. Doubts assaulted me. Hiro felt alone and isolated.

Along with Graeme and Gus, Misha was afflicted with the fewest physical ailments. As the expedition's physician, he constantly tended other people's wounds and cared for our well-being. As the one breaking trail every day amid the chaotic jumble of ice, he was the true leader and major force of Icewalk.

Many days we were reduced to crawling on hands and knees, arduously negotiating ice boulders or razor-sharp ridge backs. We fell repeatedly, dragged down by our packs and sledges, landing in a tangle of equipment, too weary to rise.

Previous to departure, I'd had a psychological work-up done on the team, similar to the one prepared for us by Dr. Elizabeth Holmes-Johnson for In the Footsteps of Scott. The report soberly concluded that our wildly differing backgrounds—something I considered Icewalk's greatest strength—was instead a killing weakness. Much better to recruit people of shared experience, the psychologists advised, to help foster team cohesion.

Wounded or not, with or without cohesiveness, we straggled on, at times strung out over a mile in the desolate landscape, each man in his own separate hell. At the eighty-fourth parallel in latitude, the sea floor rose beneath us, its upheaval creating sixty-foot pressure ridges in the surface ice.

On April 11, after twenty-three days of walking, Misha, Graeme, and Gus—the three racehorses of the expedition—were a hundred yards ahead of the rest of us. They paused atop a towering pressure ridge. We climbed, tripped, and stumbled forward to join them.

I mounted the ridge. The view to the north staggered me. I felt sick. The North Pole was melting far earlier in the season than it ever had before. As far as the eye could see, black fingers of open water blocked the way.

That's where our real troubles began.

Open Water

The Big Nail. That's an Inuit term for the North Pole, obviously coined since the days of the pre–Iron Age people's first exposure to European traders.

Well, from the looks of the open ocean atop that pressure ridge, we were not going to nail the Big Nail after all. What were we supposed to do, hop from ice floe to ice floe across the Arctic Ocean all the way to the North Pole?

That night we made camp on a floe, a house-sized frozen lozenge floating amid the black water that I had gazed across earlier in the day. All around us, only ten feet away from our tents, slabs of ice creaked and groaned, pushed upward by the sea, rising and falling ceaselessly. I lay in my sleeping bag, hearing the lap of water just below me. Far below that lay the mile-deep floor of the polar sea.

This is really happening. That had been my visceral, immediate response when I viewed the mid-April breakup of the Arctic Ocean. The environmental degradation of the poles wasn't just a good theme on which to pin an expedition. It wasn't a theory, it wasn't a ghost story used to scare children, it wasn't a *narrative*. It was frighteningly, paralyzingly real.

Standing on the shifting polar ice 350 miles from the nearest land, I had seen the future, and I knew we were in deep trouble.

That was the moment when the scales fell from my eyes. That was when I learned the lesson Antarctica had tried but failed to pound into my thick skull. That was my road-to-Damascus conversion experience.

The tent was unusually silent that night, each of us brooding on his thoughts as we lay in our sleeping bags and listened to the sloshing of open water.

"In our children's lifetime," Misha said quietly, "if our children tell their own children's friends that their grandfathers walked to the North Pole, they will get nothing but laughter. Because it will be an open sea."

I felt as though my bluff had been called. I had been gassing on about climate change, and now here it was slapping me in the face.

Wake up! I might be a slow learner, but when I finally learn something, I learn it well. That night in the tent, exhausted as I was, I lay awake, a million thoughts humming in my head.

Okay. All right. It was time to walk the talk. Camped out at one end of the earth, I reached out in my mind to the other. *Antarctica is solid ground,* I thought. *Antarctica is massive. Antarctica is key.* The continent particularly appealed to me at that moment, as I floated queasily above the Arctic Ocean. Brutal and unforgiving as Antarctica was, it seemed a haven.

The world's last pristine wilderness.

Pristine. I rolled that word around in my head. In its original English meaning from the early sixteenth century, it means "primitive, ancient," from the Latin *pristinus,* "former." Only much later, at the turn of the nineteenth century, did the sense of the word mutate into "unspoiled, untouched, pure."

Was there a place for the pristine in this world? "It is good to know," Scott wrote, "that there remain wild corners of this dreadfully civilized world."

Dreadfully civilized. I liked that. But can a modern, industrialized society afford to keep its "wild corners" pristine?

I knew that at the very moment I was lying awake on the Arctic Ocean ice, the Icewalk student team was assembling, streaming in from all parts of the world to the Eureka base on Ellesmere Island, four hundred miles to the south. When they arrived, among the other lectures, research projects, and activities, Environment Canada's Dr. Dennis Gregor, one of the world's leading research scientists on industrial contamination in the Arctic, would present the students with a disturbing discovery he had made. On the Agassiz Ice Cap only miles to the east of Eureka, his research team had detected traces of a pesticide used only in Texas, over one thousand miles away in the United States.

So in this day and age, pristine is always relative. Everything we do *here* has an impact over *there*.

Another presentation the Icewalk students were scheduled to hear was from Dr. Ian Sterling of Parks Canada. He had tested the fat of polar bears for PCBs and found outrageously elevated levels of the industrial contaminant. Sterling's research team had also detected unsafe levels of the carcinogen in the breast milk of Inuit women.

Industrial toxins tend to lodge in the fat tissues of organisms. Animals in the polar regions typically protect themselves from cold with heavy layers of fat.

Like iron filings drawn by a magnet, the large, destructive, complex molecules of the modern industrial process migrated toward the poles. Carried by winds or climbing up the food chain, the toxins served to contaminate the pristine. Both the Arctic and Antarctic were sops for our poisons.

My heart ached for the students. What a planet we had bequeathed them! A place where no corner of the nest remained unsullied. Yes, *pristine,* from the Latin word for "former." What used to be.

I didn't need Drs. Gregor and Sterling to convince me. I had witnessed the interconnectedness of the modern world with my own eyes earlier that day, looking north from a pressure-ridge vantage point and seeing open water where there should have been none. The ice in that area of the Arctic Ocean had never broken before August. That's precisely why we had scheduled Icewalk for the early polar spring. But here it was April, and the ocean was opening in front of us.

What we did down *there*—pumping hydrocarbons from our smokestacks and automobile exhaust pipes—had an impact up *here*. As I lay in the tent, conjuring up the image of black water spreading over the ice, it appeared to me to be fingers of a hand, slowly tightening, choking us off.

For the benefit of the media or in front of an audience of schoolchildren, it was extremely easy for me to mouth platitudes about climate change and the environment. They came tripping off my tongue without effort. But it was another thing entirely to confront direct evidence of global warming out on the Arctic Ocean, when it could well mean my own death.

At the very least, it meant the end of Icewalk. However inflated was my ego, I knew I wasn't Jesus. I couldn't walk on water.

Perhaps this was what I had come north to experience. Not the walk to the pole but the impossibility of it. Maybe that panoramic view of broken ocean brought me here. So I would be forced to conclude, finally, that it was time to do something more than talk.

Icewalk was ending. But I resolved that I would make a new beginning. If I managed to get off the ice and back home without being swallowed by the Arctic Ocean, I would dedicate my life to my Beardmore promise: *I will somehow do whatever I can do to protect you.*

The waters below me gurgled, and the ice slabs ten feet away groaned. *What's down beneath me now?* I hoped the ice was at least a couple of yards thick. Below that, perhaps there was the passing shadow of a bowhead whale or the impossible "sea unicorn," the narwhal. Even farther down, an American or Russian submarine running silent, running deep. When a sub surfaced in the Arctic, it blasted straight up through the ice with its conning tower. I would have liked to see that, but I hoped it wouldn't happen exactly right where we had pitched camp.

Little did I suspect that two decades on, there would be a Russian flag planted on the ocean floor beneath the North Pole, the first move in an Arctic territorial land grab. The only way to save Antarctica from being destroyed by human greed is to develop renewable resources as the main energy supply for the whole globe. It's a complete waste of time to say "save Antarctica" unless you can somehow make it useless and not at all cost-effective for the oil companies to go down there and rip the place apart.

I had a short colloquy with myself as I drifted off to sleep.

I promise I will somehow do whatever I can do to protect you. I will make an effort to link the preservation of the poles to the use of renewable energy.

The cynical voice inside me had a response. *Don't be absurd,* it whispered. *You are just one man. What can you do?*

I promise I will somehow do whatever I can do to protect you.

Pompous, overreaching bullshit, said the voice. *Go home to the United Kingdom, hit the lecture circuit, live out your life like a normal individual.*

I was at it again. You don't make promises to change your life when you are sitting at home safe in front of a warm hearth. Here I was, up against it just like I had been at the Beardmore, and I was again flinging promises to the wind. Making another vow to the gods of the ice so they wouldn't bloody kill me.

I promise I will somehow do whatever I can do to protect you.

Right, dear chap—I'll believe it when I see it.

I promise . . .

The Big Nail

I awoke the next morning to find Misha preparing the kit for another day on the ice.

"What are we doing?" I asked him.

"Forward, always forward," he said in his chopped, accented English, working on his backpack.

"What about the water?"

"We go around." He looked over at me. "We know we encounter open leads soon or later."

"This isn't normal, is it?" I asked.

"No, not normal," Misha said. "Weeks and months before it should."

Open leads. *Leads* was the term for those clutching fingers of black water. As the ice shifted and broke apart, it opened gaps, long and thin, where the sea showed itself. When I looked into the choppy waters of an open lead, I was looking at mile-deep ocean.

Without Misha we would have been sunk, literally. His approach through the maze of open leads was simple. Follow alongside one until it narrowed enough for us to leap across it. Continue forward until we reached the next lead, then repeat the process. Our route to the pole was thus in no way a straight line. More like crinkum-crankum.

Oftentimes I'd be following along Misha's track on one side of an open lead and see him chugging up on the other side, eight or ten feet away and going in the opposite direction. We'd raise a ski pole in wordless salute and continue on our dogged way. Graeme estimated that such back-and-forth traveling added a fourth more distance to our walk to the pole.

As crossing an ice crevasse was in Antarctica, every lead crossing in the Arctic was an adventure. We crossed leads by balancing on ice floes. We crossed leads by throwing chunks of snow into them, building up a sloppy, mushy surface until it was "safe" to pass over. We crossed leads on skims of ice inches thick. We raced across leads as they broke open and widened, blocking any retreat.

While we crossed a lead, the ocean made an odd hissing noise below us, and sea smoke rose from it. I thought it looked and sounded angry at our intrusion.

In the north there were none of Gareth's sea leopards to menace us from below. The only real animal threat were polar bears. We carried along a .270 rifle as protection. This was their element—*arctic* means "place of the bear," referring actually to the constellation Ursa Major but just as well applied to the great beasts. In the kind of whiteout conditions we often faced, polar bears have been observed advancing on prey using their paws to cover their noses, in order to camouflage the only prominent black part of their bodies.

We never met up with bear, but it soon became evident that our tortuous path through the broken ice meant we were probably not going to make our ultimate goal. The polar sea was breaking up faster than we could traverse it. Every day was an agonizing slog. We became too weary to be afraid of open water.

When we weren't encountering leads, we hit pressure ridges. We fell behind in our pace. On a day of tremendous effort in a howling ground blizzard, we discovered that we had actually moved minus one mile toward the pole, because the ice beneath us was drifting south while we struggled north.

On my mind were Scott, freezing to death when only eleven miles away from a supply depot, and Shackleton, who famously turned back when he was less than a hundred miles from being the first man at the South Pole. Growing up, I couldn't understand. Why didn't Scott just buckle down and do another eleven miles? How could Shackleton stop a mere ninety-seven miles away from immortality?

But on Icewalk, I came to understand how those early explorers fell short. There were times when the impossibility of taking another step overwhelmed me, and the idea of walking another eleven miles was as remote a possibility as flapping my arms and flying to the moon.

I remember looking up at an airliner, a tiny, winking dot in the sky, flying the polar route from America to Europe. Why wasn't I up there?

I visualized the flight attendant, a beautiful woman, leaning over me and serving me hot coffee. I smelled her perfume, saw the steam rise off the cup she offered me, and felt the warmth of it in my hands. The image was so real it brought tears to my eyes.

Daryl was in increasingly bad shape. His heel suppurated. Misha had glimpsed raw bone while dressing it one evening. Daryl walked in agony every day. He had to discover somehow that it wasn't his body that was going to get him to the pole but his mind. He had to find the will to go on.

I marched with him at the back of the team. "I'm going to make it," he told me. "People live in more pain than this every day."

It was an unspoken concern with the whole team: Should Daryl be evacuated?

I balked at the idea. I desperately wanted to reach our goal together. If eight people from seven countries could not work together effectively as a team, what hope was there for a world with some 195 countries, many of them at odds?

It was not Daryl but I who almost crippled the expedition. Scaling a pressure ridge, I bent my weight on one of my ski poles, hauling myself forward. With a sense of dread, I felt the pole give way beneath me. I fell, and when I came up with the pole, it proved to be broken.

A minor equipment mishap. But nothing was minor in a pole walk. There was no way to haul a sledge in these conditions without the support of two sturdy poles. It was as though I had come up lame. We carried extra poles, but we'd been breaking ones steadily. I had given out the last spare to Daryl a few days before.

I struggled forward to join the team. Misha and I conferred. Could we repair the pole? It looked doubtful. While not broken through, the metal staff was bent and torn, exposing the hollow core. What we needed was some kind of shim, a small piece of wood or metal to insert into the hollow shaft in order to keep it straight and strong. There was no handy hardware store to supply a suitable fix. The nearest one was probably, oh, a thousand miles to the south. A broken pole meant that I was stranded, or at the very least that I would slow down a trek that was already falling behind schedule.

I envisioned the black fingers of water reaching for us, the ocean opening, a humiliating rescue.

"Robert," Misha said. "Do you have that wood from Captain Scott's ship?"

Once again, Emma the clairvoyant had come through. I retrieved the small section of timber from *Discovery* that she had presented me as a good-luck gift. Rupert whittled it down. Shavings of an oaken ship, wood that had been fashioned nine decades earlier in a Dundee shipyard for Britain's first major exploration of the Antarctic, fell onto the Arctic ice.

We wedged our little memento of Robert F. Scott into the shaft, the pole held, and I was back in business.

But such incidents meant we were still falling behind the pace we needed to succeed. Equipment failures, injuries, the impossibly tortured back-and-forth path we were taking to the pole—it looked as though the sea would open beneath us long before we would attain our goal.

It was Hiro who solved our predicament. Because of our language difficulties, he was the most isolated member of the expedition. The wounded Daryl was tireless in helping him with English, but it was slow going. I remember how Hiro's face lit up when I periodically dug into my pack and retrieved the gifts I had been saving for him—copies of a month-old Japanese newspaper.

As we dodged leads and dashed across spongy, crumbling ice, Hiro skied up beside me. "The day," he said, "is not the day."

Uh-huh. Not for the first time I regretted not working harder on my Japanese. I nodded and made to ski on.

"No," Hiro said. He pointed his ski pole at the Arctic sun. "You know? The day!"

Like a bolt from the blue it hit me. Hiro was making an obvious point. Our expedition "day" had been modeled on a temperate-zone schedule. We skied for eight hours and then knocked off at night—even though there was no darkness to the Arctic night. Up here, the day extended for twenty-four hours. The day was not the day. We could utilize as many of those hours as our stamina allowed. We weren't limited to an eight-hour day.

From that point on, we powered forward for ten, twelve, sometimes fifteen hours a day, racking up the miles, racing the ice. I remember our first six-hundred-minute day, with ten hourlong marches.
Misha beamed. "Three weeks ago, I would not have believed this,"
he said.

During those desperate days and nights, I thought not about Scott,
not about Amundsen, but about Ernest Shackleton.

Of my three Antarctic heroes, Shackleton had the spottiest list of
achievements. When Scott took him along on the 1903 *Discovery* expedition, he was stricken by scurvy and had to be sledged back to base
from an aborted attempt at the pole.

Then, on his own *Nimrod* expedition of 1907–9, Shackleton
achieved "farthest south" but stopped that fateful ninety-seven miles
from the South Pole. He had realized that if he pushed on, he and all his
expedition mates would perish. "A live donkey is better than a dead
lion, isn't it?" Shackleton said to his wife in explaining his decision to
turn back.

In his third and final Antarctic expedition in 1914, Shackleton's
failure was even more complete: He never even made it to the continent. The ice pack gripped and crushed his ship, *Endurance,* and he
and his men were marooned on the ice of the Weddell Sea.

It was his biggest failure, and his most celebrated success. The story
of Shackleton's leadership of this shipwrecked expedition is one of the
truly thrilling tales in the annals of exploration. He and his crew drifted
on ice floes and in lifeboats to desolate Elephant Island. From there,
Shackleton and five others, including his superb navigator Frank
Arthur Worsley, crossed the South Atlantic Ocean. In an epic eight-
hundred-mile open-boat journey that stands as one of the most astonishing feats in the history of sailing, they made land and marched to a
whaling station on South Georgia Island, eventually accomplishing the
rescue of all their expedition mates.

Wrote Edmund Hillary: "For scientific discovery, give me Scott, for
speed and efficiency of travel, give me Amundsen, but when disaster
strikes and all hope is gone, get down on your knees and pray for
Shackleton."

On Icewalk, I got down on my knees and prayed for Shackleton. The melting Arctic Ocean made disaster a constant, looming companion on the last third of the expedition. Exhausted, stumbling on hands and knees over ice ridges as tall as a two-story building, or skirting black seawater leads, I reduced my world once again to a single, endlessly repeated word, my mantra.

"Forward, forward, forward," I muttered to myself, timing each syllable to another step.

At 3:30 A.M. local time on May 14, 1989, the Icewalk expedition made the North Pole. We ascertained our position at 90 degrees north via tracking satellite. Rupert had navigated us to within 438 meters of true north, matching the accomplishment of Roger's near-dead-on reckoning in the south three years before.

On a cairn of ice boulders in the middle of a barren expanse of snow, we raised the proud blue flag of the United Nations.

Canary

It is all very well to make a promise and quite another thing to figure out how to fulfill it. I could promise anything, even, as President Kennedy did, to fly to the moon. That's the easy part. The toils of NASA came afterward.

Slowly the new, changed world I had discovered out on the disintegrating ice of the Arctic came into focus. Political upheavals mirrored environmental ones. Misha had left on Icewalk a Soviet and returned a Russian. Arved had left a West German and came back a German. Change was the only constant.

I had promised to try to protect Antarctica. Protect her from what? What were the threats? I held the whole continent in my mind as an enormous, inaccessible, unsullied realm. What would change that? What could possibly crash in upon it and violate it? What would be most likely to bring it to ruin?

I put on my prognosticator's cap and peered ahead in time. The answer came fairly readily. The world's supply of precious metals and rare minerals was diminishing, used up by the voracious industrial appetites of the modern age. Bauxite, for example, the essential aluminum ore, represented a finite resource. Add mining to the specter of oil exploration, and Antarctica appeared vulnerable to all sorts of human mischief.

But Antarctica was protected, wasn't she? I dug into the Antarctic Treaty System (ATS) and read and reread the newly minted Madrid Protocol, officially the Protocol on Environmental Protection to the Antarctic Treaty, negotiated in Spain by the signatory nations of the ATS in 1991.

Article 7 of the protocol seemed fairly airtight. "Any activity relating to mineral resources, other than scientific research, shall be prohibited." I also liked the language of Article 3:

> The protection of the Antarctic environment and dependent and associated ecosystems and the intrinsic value of

> Antarctica, including its wilderness and aesthetic values
> and its value as an area for the conduct of scientific re-
> search, in particular research essential to understanding
> the global environment, shall be fundamental considera-
> tions in the planning and conduct of all activities in the
> Antarctic Treaty area.

The treaty would effectively preserve Antarctica from commercial exploitation. But as Tom Waits sang, "The large print giveth, and the small print taketh away." I noticed a disturbing clause in the Madrid Protocol. It would remain in force for only fifty years. At the end of this period, its operations could be amended to allow for all sorts of mischief, including mineral exploitation.

Here's the precise language: "If, after the expiration of 50 years from the date of entry into force of this Protocol, any of the Antarctic Treaty Consultative Parties so requests by a communication addressed to the Depositary, a conference shall be held as soon as practicable to review the operation of this Protocol."

Fifty years. Fifty years from 1991 would place Antarctica in renewed jeopardy in the year . . . 2041. As I peered into my crystal ball I could very easily see the world changing over the course of the next half century. It could be a much different place than it was in 1991.

I imagined the future world as a well-fouled bird's nest, crowded with the cheeping, gaping mouths of baby birds. More petroleum, please. More bauxite, more platinum, more copper.

Copper, lead, zinc, gold, and silver have been found on the Antarctic Peninsula. Geologists have located deposits of low-grade coal in the Transantarctic Mountains, and there are some indications of the presence of chromium and platinum in the Pensacola Mountains, near the Ronne and Filchner ice shelves. Surveys turned up iron ore in East Antarctica.

The real lure, though, will be oil. Sedimentary basins in the Ross and Weddell seas, and near Prydz Bay on the little-visited Ingrid Christensen Coast in the east, might be enough to put a gleam in the eye of a predatory petroleum geologist. (Sedimentary basins are to petroleum geologists what blood banks are to vampires.) So far, however, no one has had the temerity—or the funding—to conduct extensive basin analysis in any area of the continent.

I keep hearing an echo of the corpulent merchant in *Scott of the Antarctic*. "What are the prospects of trade between here and Antarctica? Is there anything that I can buy and sell?"

Well, it turns out that possibly, probably, there is. And in a future era of diminishing resources, the cost-benefit ratio might tip in favor of going down to get it.

How could I help prevent commodity-hungry eyes from turning south and contemplating Antarctica as an untapped source for increasingly rare precious minerals?

One way I could do it was to fix the year 2041 in the world's mind as a decision year for Antarctica. If as many policy makers as possible could be made aware of 2041, perhaps the operations of the Madrid Protocol could be extended and the continent preserved from being date-raped by mining and energy interests. Elevate 2041 to an idea, make it iconic, transform it into a meme.

That, anyway, was my thought. Then I considered the most important element of the equation: the "policy makers." Who would be the people who would decide to re-up the Madrid Protocol in 2041? I realized that in that year I would be . . . eighty-five years old. If I lasted that long, most of my decisions would involve how many times in the night I would have to get up to pee. Young people *now* would be the decision makers *then*, in 2041. They would form the core group of leaders who would be forty, fifty, or sixty years old in that watershed year for Antarctica. When I was speaking to a young person, I was speaking to someone who might be a power, a person of authority, in 2041.

So, yes, the next link in my chain of thought: Reach out to people who are under age thirty, under age twenty, under ten, for pity's sake. Get them interested in preserving the last great wilderness on the planet. Inspire them with the 2041 meme. So that by the time the treaty operations came up for review, the result would be virtually automatic.

It wasn't much. It was only a small piece of the global environmental puzzle. But Antarctica is a vital piece. Hydrologists estimate that the

Antarctic ice cap locks up an incredible 90 percent of the world's ice and 70 percent of its fresh water. Melt even a small portion of that, and say good-bye to New York City, London, Hong Kong, Cape Town, Rio.

I've always believed that the idea of "saving" the earth is an example of absurd hubris. The planet will still be here even if humankind screws it up to a degree that renders the place uninhabitable. Perhaps, as in the film *Wall-E,* the durable cockroach will survive. What "save the planet" and "save the earth" really mean is the preservation of a human-friendly home. Save our planet. Save our Earth.

But "save Antarctica"—that's a two-word phrase that can be taken quite literally. Beneath its overbearing ice cap, the continent is a tattered archipelago, with an uneven coastline and fjordlike fingers reaching into the interior.

Jack up global temperatures high enough, melt all the ice in the Antarctic ice cap, and the place would disappear underwater. It might take centuries, but if we continue on the path we presently tread, the Beardmore will be gone and Scott's hut will be submerged.

The history of humans in Antarctica is bittersweet for me. It's a short history, the shortest of any continent. There is no record of anyone seeing Antarctica before January 27, 1820, when Fabian von Bellingshausen's Russian expedition first sighted the coastline through Neptune's Window on Deception Island off the Antarctic Peninsula.

Soon afterward the sealers poured south with their clubs, guns, and knives, seeking profit. The year 1821–22 represented a seal holocaust, when whole populations of the creatures were hunted to commercial extinction and a high-water mark of 164 ships pursued prey worldwide (ten years later there was only one sealing vessel registered with Lloyd's of London). On Deception Island whalers established bases and factory ships that facilitated the slaughter of millions of seals and more than one million whales in the course of two decades.

"Commercial extinction." What an ugly little phrase that is, what a harrowing concept. It means that creatures have been harvested to such a degree that the search for the remaining few victims no longer makes commercial sense. The calculus of profit is unassailable. When the hunt costs more than the harvest brings in, it's time to quit.

The sole virtue of commercial extinction is that it represents a slightly preferable state of affairs when compared with actual extinction. Many species, the American bison most notably, have been saved from actual extinction by commercial extinction.

Only after the sealers had hunted out their prey did the Heroic Age of Antarctic Exploration really begin in earnest. Then came the race to the pole, the various transcontinental crossings and flyovers, and the geopolitical jostling that ultimately led to the first elements of the Antarctica Treaty System on December 1, 1959.

The world's polar regions are justly cited as the planetary equivalent of the canary in the coal mine. You see that metaphor being bruited about a lot without really thinking about what it means. Sting wrote a song called "Canary in a Coal Mine." The phrase brings up more than 300,000 Google hits.

The practice began in the Westphalian coal mines of Germany. In the nineteenth and early twentieth centuries, miners used to take along caged canaries down into the shafts, because the birds would detect lethal gases in the mines before the poisoned air would affect the workers. Canaries are extremely sensitive. Open a permanent marking pen such as a Magic Marker next to one, and the little bird will pass out from the fumes. The birds carried into the mines alerted workers to danger by fainting or simply keeling over and dying. They functioned as a biological early-warning system.

In the Westphalia mines, the birds used were often tuneful Harz rollers, bred to sing in a fantastic array of vocalizations. The miners received the collateral benefit of canary Muzak, as the birds whistled while the men worked.

Extending that metaphor to the polar regions, we can see that Antarctica would have to die before we would wake up to the threat to the whole planet. The polar regions make up our planet's early-warning system. Peering into the Antarctic environment is like peering into the future. What happens there now will affect the rest of the globe in the years to come.

So what I wanted to say with the 2041 effort, more or less, was "Keep the canary alive." But the whole initiative wasn't really about the canary. It was about the coal mine. It was not about the bottom of the world as much as about the world itself.

In another sense, the world's canary could literally turn into the world's coal mine, if mining were to be allowed on the continent. Journey through the mountains of West Virginia to see what irresponsible exploitation has done to that environment. Now imagine that writ large, across a continent twice the size of Australia, more fragile than a coral reef.

The idea was underscored in Brussels in 1990, when I attended ceremonies to receive the Global 500 award from the United Nations Environment Program. That same year Margaret Thatcher was also receiving the award, officially called the Roll of Honor for Environmental Achievement. According to the program, "The individuals and organizations who join the prestigious ranks of global laureates have, in their own distinctive manner, influenced the destiny of life on earth as active participating members of the community."

It was purely honorary. The Global 500 award, plus the correct fare, would get me on the subway. Such an accolade was like the proverbial kiss from your sister—nice and loving, but it wasn't going to get you anywhere. The occasion was most memorable to me because I got to spend a good deal of time with Jacques Cousteau.

Through Sir Peter Scott, I had met the great oceanographer and environmentalist before, briefly, at UNESCO headquarters. It is difficult to describe to young people what Jacques-Yves Cousteau meant to the world in the last half of the twentieth century. He was the first conservationist celebrity since John Muir. But mention the name to someone under twenty, and you will most likely get a blank stare.

In Brussels, we bonded. Sir Peter Scott had just died. Cousteau was eager to honor the man who had done so much for wildlife and conservation. I think he saw mentoring me as a way to do that. And they both had specifically asked me to be sure to clear up our expedition rubbish from Antarctica.

Amid the glitz and glamour of the Global 500 celebration at a Belgian *palais,* Cousteau and I were paired at a photo opportunity, the ceremonial cutting of a large cake. It felt ridiculous, but Jacques's smooth Continental manners soothed my nervousness. He was always an elegant man, forever dressed in a turtleneck sweater and a blue linen blazer.

Afterward, we talked. I was still nervous. Cousteau was one of my heroes. Simply for being the codeveloper of the Aqua-Lung underwater-breathing apparatus, he deserved awe. Oftentimes, as a teenager, I had tried to imagine the courage it took to conduct underwater testing of the first Aqua-Lung.

His earlier advice to me had paid off, I told him. Icewalk was my way of using a short-term mission to advance a long-term goal. I mentioned my feelings about protection of the Antarctic and the significance of the year 2041.

He looked troubled. "I am unsure that we have fifty years," he said in his perfect, French-accented English.

The hauntedness in Jacques Cousteau's eyes that day stayed with me. I have encountered the same look repeatedly over the years, speaking to other scientists and researchers who have studied the effects of environmental degradation and climate change.

Along with the polar regions, the ocean is another kind of canary in a coal mine. It registers human recklessness and profligacy as surely as a barometer measures air pressure. Cousteau had seen what was happening firsthand, and the haunted, hunted look in his eyes reflected it.

After my experience on Icewalk, and encouraged by the Global 500 award and Jacques Cousteau, I resolved to somehow bring my concerns about the polar environment to the wider world. At one of the most important gatherings of political leaders in the twentieth century, I had an opportunity to do just that.

Rio

Come on, world leaders, cheer up!"
Rio de Janeiro, 1992. I stood in front of an audience of 108 presidents and prime ministers at the world's first Earth Summit—or at least the first one in twenty years, since the acrimonious Conference on the Human Environment met in Stockholm in June 1972.

I looked out at the crowd I was addressing. Everyone appeared very glum. From the looks on people's faces, no one seemed to want to be there. There was John Major from the United Kingdom. Helmut Kohl from Germany. The first President Bush, George H. W., in the middle of what would be a losing reelection campaign, who managed to alienate the conference participants by expressing an unwillingness to get behind climate-change initiatives.

I myself did want to be there. I thought it was great.

It helped that I had been invited to the Earth Summit by none other than the Iron Maiden herself, Margaret Thatcher. After my return from Icewalk in May 1989, the prime minister summoned me to No. 10 Downing Street. The photo of the occasion shows me with my totally flummoxed just-back-from-the-pole look of disorientation, wearing shorts while the more formally attired Maggie smiles stiffly beside me.

"Rob, you must go to Rio," Margaret Thatcher had said to me.

To which I had thought, *Wow, what a fantastic, caring lady. She's sending me off on a much-needed vacation. Rio sounds nice. They've got good beaches there.*

The reality only dawned a year later, when I received a thick, official envelope in the mail, a formal government invite to the Earth Summit. John Major had supplanted Thatcher as prime minister by then.

Earth Summit, a.k.a. the United Nations Conference on Environment and Development, a.k.a. Eco '92, a.k.a. the Rio Summit. One-hundred seventy-two nations of the world present and accounted for.

The Rio Central conference hall, all tubular steel and glass windows, was spiffed up and packed to the rafters.

At a contemporaneous Global Forum, 2,400 representatives and 17,000 supporters from hundreds of nongovernmental organizations such as Greenpeace and the World Wildlife Fund turned out. This was the unofficial half of the conference, and a little more colorful than the sober goings-on in the conference hall. A Viking ship named *Gaia*, after the Greek earth goddess, pulled up dockside, bearing thousands of letters from children, each pledging to protect the environment.

The gathering was unprecedented for a U.N. conference, both in terms of size and the scale of its concerns. Officially, I was there in my capacity as U.N. Goodwill Ambassador with Special Responsibility for Youth, and as special envoy to the director general of UNESCO. *Special* seemed to be the operative word here.

Once again, as with the Global 500 award, those titles were purely honorary. But they lent a modicum of weight to my 2041 push, which was, back then, just being born.

My mother put it into perspective. "You know, Robert," Em told me, "you've been John the Baptist, a voice crying in the wilderness, for so long, talking about global warming, the environment, and no one would listen. Now you've come in from the wilderness. Isn't it fantastic that people are finally getting up to speed?"

I was up in front of the Rio summit primarily because of Jacques Cousteau and my Global 500 award. I had five minutes on a Sunday morning to address the assembled prime ministers, presidents, and potentates.

"Come on, world leaders, cheer up!"

Nothing. The audience was as cold as the Beardmore winds. My jokes fell flat. It was the toughest crowd of my life. My face went red and my mouth went dry. It was a very long five minutes. I thought perhaps I was getting lost in translation. Here it was, this was my chance, and I was blowing it.

Incredibly, no one seemed to notice. Drenched in flop sweat, I came off the stage at Rio Central convinced that I had set the environmental movement back years. But several people congratulated me. There were smiles and nods. I realized that a United Nations conference was not Broadway or the West End. You didn't need to *kill*, in show-biz parlance. Nikita Khrushchev's shoe-banging episode notwithstanding, the U.N. was not exactly known for heights of oratory.

There to congratulate me after my speech was a good friend, Jonathon Porritt, the environmental campaigner I've come to respect the most. Just days before, he and I had been pinning up letters from young people on Rio's symbolic "Tree of Life" in front of the conference center. He would become Lord Porritt, a lifelong ally, a mover and shaker in the Green Party and the Friends of the Earth group.

But I was disappointed in myself, because I believed Rio represented the kind of international cooperation that was essential for the goals of 2041 to be achieved. The initiatives begun at Earth Summit eventually led directly to the international Kyoto Protocol on greenhouse-gas emissions.

That major achievement had a darker, flip side in the politicization of the whole question of climate control. First the U.S. Congress and then the second President Bush, George W., blocked the American ratification of the Kyoto Protocol. The world's leading producer of greenhouse gases refused to sign on to the treaty designed to control them.

Rio cemented my conviction that for me to fulfill my promise to do whatever I could do to protect Antarctica, I would have to think globally rather just in terms of the continent itself. We could build a Great Wall around Antarctica, and it would be useless if climate change generated in other parts of the world continued to degrade the polar environment.

"Think global, act local." That somewhat ungrammatical phrase became the motto of the Earth Summit. But Rio also gave birth to a concept that I took as one of the key ideals of my life: sustainability.

The word had been knocked around long before that, of course, but "sustainable development" was the hallmark phrase at the conference, which, like all U.N. hallmark phrases, generated a commission to study it, foster it, and report to the General Assembly on progress toward the goal.

I've thought a lot about that word since the Earth Summit. There is

an unassailable logic to sustainability, but there is also an almost in-
stinctual human resistance to it.

Try it yourself. Drop *sustainable* or *sustainability* into conversation.
Sleep instantly overtakes your listeners, a glazed look comes across their
faces, their eyes go blank. I think this is because sustainability carries
within it a concomitant idea of responsibility. It requires, for each and
every act, an awareness of effects through time and space.

Walking out of the room and leaving the lights ablaze: a casual,
quotidian action, seemingly minor and meaningless. The idea of sustainability requires us to expand that simple act in time and space, examine it in terms of how it may affect Rio, Calcutta, or Antarctica, next
week, next year, far into the future. In an interconnected world, *nothing* is minor, nothing is meaningless.

On the ice above the Arctic Ocean, I had seen with my own eyes the
effect of millions of people leaving the lights on. In the Arctic and the
Antarctic, our casual, mindless acts—driving to the store to pick up a
quart of milk, purchasing a big SUV, burning coal at a power plant—
first showed up as meaningful.

Sustainability brought baggage with it, the quite tiresome require-
ment to analyze what previously had required no analysis. It challenged
our attention deficit disorder. It was a gigantic pain in the arse. This, I
believed, was the source of the knee-jerk resistance to the concept that
I encountered wherever I went.

It is the problem of "should" that has plagued the environmental movement from the beginning. We should do this, we should
do that. The environmentalist is a natural scold, a nudge, a finger
shaker. But negativity has never motivated anyone. We needed somehow to progress from a what-should-be-done mentality to a can-do
determination.

My own contribution to the already teetering juggernaut of sustainability was an elaboration of the idea. Thinking about it after all the
fine, high action of Rio—and throughout the fractious political battles
there, too—I realized what we needed was sustainable leadership. I had
an awful feeling, during those days at Earth Summit, that there was
going to be a lot of talk, then everyone would go back home and the
momentum would slowly dissipate.

To address a global, long-term goal such as halting human-induced climate change, we needed to be able to sustain an initiative. We needed to be in it for the long run. We couldn't hold a conference and then let our attention wander.

By focusing on the year 2041 I had already recognized this—made a decades-long commitment toward the goal of protecting and preserving Antarctica. And I had witnessed sustainable leadership over the course of my expeditions. What Roger Mear and Misha Malakhov did was sustain an effort, day after gruelling day. To take one's eyes off the ball in the polar regions meant failure or, in the extreme case, death.

Without really being aware of it, I had become a student of sustainable leadership. I had watched Roger and Misha. I had seen, over the years of mounting In the Footsteps of Scott and Icewalk, what worked and what did not. I had been able to sustain both expeditions through adversity, reversals, and serial mishaps.

Hidden within the idea of sustainability is an interesting question. What does it take to make a team successful?

Consider a group of people in pursuit of a goal. It could be any goal—a sports championship, a profitable quarter, a walk to the pole. What elements come together to make one team successful where others are not? Out of the multitude of variables in any human interaction, which are the vital ones? What helps? What doesn't? What sustains us?

Sustain—from the Latin for "under" and "to hold." The underpinning, the foundation. I hate to state it so clearly, at the risk of making anyone's eyes roll up into their head and the snores kick in, but the single concept of sustainability holds the key. It certainly did for me on Icewalk and In the Footsteps of Scott. And it would, I believed, prove crucial to the eventual success of 2041.

At Rio in 1992, as Goodwill Ambassador and in the spirit of "Think global, act local," I signed on with world leaders to create two initiatives for young people. A global mission and a local mission. The global mission would address some worldwide issue or problem. The local mission would be based in Johannesburg, South Africa, where in 2002 another Earth Summit would be held.

After my flat-footed talk in front of the whole conference, UNEP chief Dr. Mustafa Tolba ushered me into an inner-sanctum gathering. *Ah,* I thought, *the smoke-filled room. This is where the real work gets done.* I felt totally out of place among the power suits. I saw President Bush on one side of the room with Helmut Kohl.

Dr. Tolba introduced me to John Major. He mentioned that I had announced two initiatives to work with youth. "One global, one local," Tolba said, nodding and smiling. Major nodded back. I just stood there, silent and uncomfortable, somewhat like a waiter hoping for a tip.

The prime minister looked around with a distracted expression. *Oh, Swan, are you still here?* "Carry on," he finally said. "We'll see you at the next summit."

When the leader of Britain grips your hand and looks into your eyes, and standing around behind him is the American president and the chancellor of Germany, you only really have one course of action open to you.

You say yes. Like the title of the BBC television program, you say, "Yes, Prime Minister."

But really, what I was thinking was, *Now I'm stuck with it. One day I'll learn to keep my mouth shut.* My debts hadn't gone away. And I had just taken on a huge new commitment.

I stood around at the elite gathering some more, a fish out of water, hoping someone would come up to me and say, "Here's some money."

No such luck. I now had a global and local mission to come up with. But first I had to fail, and fail big, crashing to the earth and aborting a polar mission for the first time in my life.

Barney

Since In the Footsteps of Scott and through Icewalk and the Rio Summit, I've led two lives, two existences that balanced on top of each other, to use Bob Dylan's indelible image, "like a mattress on a bottle of wine."

One life was my everyday, ordinary one. Waking and sleeping and eating and drinking. My other life—and this one was the mattress—was my pursuit of expeditions and fund-raising and lecturing.

Raising money to walk to the poles is more difficult than actually walking to the poles. I have always been the biggest sponsor of what I do. I've repeatedly spent myself broke (and then some) to mount expeditions. This kind of questing profligacy doesn't sit well within the framework of a private, domestic existence. The two lives existed in uneasy relation to each other. Attempting to sustain my leadership in a professional sense rendered me unable to sustain private, day-to-day relationships with anything that resembled consistency or sanity. I sacrificed one for the other.

Being stuck on mile eighteen made for good progress in my professional life. I was all-out all the time. But it was proving hell on my private, ordinary existence. Too often I found myself literally balancing my public life on a private bottle of wine. I consumed barrels of it, red or white, it didn't matter.

"How do you like your brandy, sir?" asks Charles Waldron in the movie The Big Sleep.

"In a glass," says Humphrey Bogart.

That was me. I felt as though I had been living in a twelve-year storm, from the time I left university in 1980 until Rio in 1992. I had two contradictory questions plaguing me.

How do I keep it going?

How do I make it stop?

How do I keep juggling expeditions and debts and fund-raising and lectures and public appearances? And how can I let the balls drop and walk away? I continually felt exhilarated and exhausted at the same time.

Something else happened in Rio. I reconnected with a friend I had known a long while before, when she lived in London. Nicky Begg was a longtime environmental advocate who came to the Rio Summit at the same time I did. We got reacquainted over, yes, a bottle of wine, and in the following year we were married.

I thought I needed an anchor in the storm, something to fasten me down to the here and now. Nicky had the courage and the love to become that anchor.

Slowly, I became aware of a dogged undertow in my life. I was beginning to feel . . . well, *curse* is too strong a word, but there is a haunted quality to people who have spent a lot of time in Antarctica. The continent's desolation works its way inside you.

The comfort of a temperate-zone existence can act as a buffer, separating the self from the harsh realities of existence. Warm breezes, gentle rains, the quality of mercy—our species is tailor-made for certain latitudes.

Antarctica strips that buffer away. Like the splinter of mirror that lodges in Kay's eye in the Hans Christian Andersen tale "The Snow Queen," a small shard of Antarctica can permanently affect your vision. The existential challenge of the place remains with you long after you've left. You tend toward broodiness.

At least, I did. I tried to stay away. For a long time, I sought to mount expeditions in warmer climes, far away from the poles. I spent time with Nicky and tried to decelerate from mile eighteen, seeking out the normal, the humdrum. But something was wrong. I continued to drink heavily.

Then, in 1994, everything changed. My son, Barney, was born. Barnaby Swan, named after one of my best friends from Durham University, Barnaby Gaston, and because our home in England was near Barnard Castle, one of my favorites places in the world, a grand, twelfth-century ruin on the north bank of the River Tees.

Trailing clouds of glory did he come, and because of Barney I felt a fierce, protective love motivating me. The birth of my son made me realize, in a way I never had before, what I was doing all this for.

Barney also made me realize that I might have a drinking problem. I wanted to be the best I could for him, and my crapulous behavior got in the way of that. I knew I had alcoholism in my background. The traditional Brit approach was to suffer such issues in silence. But I didn't think I could make it alone.

With the support of Nicky, I reached out for help. The professional I met, an extremely smart counselor named Mark Fish, would not even listen to me the first day. He held up his hand to stop me before I opened my mouth. "Robert, before we meet, I want you to do something," he said. "There's a bookstore around the corner. I want you to go in there, buy a copy of the 1995 edition of *The Guinness Book of Records*."

I looked at him. What, right now?

"Go ahead," he said. "Buy a copy, and look at page 178."

I left, wondering if he was going to charge me for a whole session. But I went to the bookstore and opened up *Guinness* to the suggested page. I already suspected what I would find there. "First person to walk to both North and South Poles: Robert Swan."

The next time I met with Mark Fish, he asked me if I had done what he had told me to do. I said that I had.

"So?" he said.

"So?" I answered back, a little nonplussed. If this is what therapy was, I wasn't getting it.

"So, Robert, you did it."

"Yes," I said.

"No, no," he said, emphasizing the words. *"You did it."*

"Okay," I said, still not getting it.

"You did it," Mark Fish said again.

"Right," I said.

"So why are you behaving as though you are still doing it?"

From that point on I began the process of getting rid of that particular crutch, the crutch of alcohol. Fish had unerringly zeroed in on my compulsion with mile eighteen, with always acting like I had to best an opponent or beat a challenge. He helped free me.

I got help too late to save my marriage, but getting sober did manage to save the friendship between Nicky and me. She moved to Australia and remarried, but as the father of our child, I remained close to her. Newly sober, I swore an oath to myself to be the best dad I could possibly be to Barney, wherever I was in the world, and wherever he was.

I also had help from two friends who have been stalwarts in my life. I've known Samantha since the Icewalk days and Lavinia since long before that, when I was twelve. They are loving, supportive friends, tremendously important because they both try to make me stop and think when I'm so wrapped up in the moment that I can't see beyond my own nose.

"Rob, just think about what you've done over the years," Lavinia will say if I'm whining about this or that. "Quit pushing on for just a second. Look what you've accomplished for the team."

Samantha and Lavinia help me keep perspective. It's vitally important for everyone, not just leaders but leaders most especially, to make sure to celebrate successes and mark achievements. Otherwise you become like I was, stuck on mile eighteen and trying to drink my way out.

The BBC folks had me on a talk show to speak out about my struggles with addiction. I got more responses—more letters, more e-mails—from that single appearance than I ever got from interviews about Footsteps of Scott or Icewalk.

"It's terribly important if you have this problem to be supported when you ask for help," I said, looking soberly into the camera lens. Not exactly a revelatory statement today, but back then in the United Kingdom, it was something people needed to be reminded about. The British mind-set was such that you had to be a wimp or a failure to reach out. My essential message: If a macho-man pole walker can do it, anyone can.

Barney's birth put me on the path. But Mark Fish taught me that I couldn't do this for other people. I had to do this for myself.

It's like they tell you every time you fly. Put your oxygen mask on yourself first. Then you can look around in order to help others.

One Step Beyond

In an odd paradox, the charge of inspiration I got from Barney's birth actually worked to send me away from him. Determined to work for Future Generations, capital *F*, capital *G*, I traveled incessantly, leaving my baby boy at home.

Sir Ran Fiennes, the British polar explorer, once defined his role stiffly but succinctly: "Expedition leadership was my chosen way of making a living." I was embarrassed to face this bald truth, but there didn't seem to be an alternative. In a way, expedition leadership chose me. It was what I was good at, what I knew, what I did.

After Rio I really did cast around for temperate or tropical expedition ideas. I dabbled with taking young people to different parts of the world. *Wouldn't it be great,* I thought, *to be plunked down in the desert with a solar panel?* Someplace warmer, anyway, than Antarctica, my bête noire, or actually my bête blanc. I was sick to death of the polar regions. I went through all kinds of turmoil, trying to come up with a mission that answered the siren call of John Major. I tried on the idea of "Earth Rally" or "Rally for the Future." Two cars, driven all around the world, in competition to see which one would use the least fuel.

Or how about "Earth Walk"? A series of walks with groups of young people, visiting the endangered parts of the world. The *warm,* endangered parts of the world. Borneo, the Amazon basin, sub-Sahara Africa. Plenty of areas to choose from.

I circled the question. If I was advocating for the Antarctic, did it necessarily mean I had to go there?

But I was kidding myself. The pull-challenge-love-hate of the south proved too strong. There was nothing for it. An idea I had been mulling over ever since Footsteps of Scott gelled in my mind.

I had done half of Antarctica, from the coast to the pole. How about doing the other half, from the pole to the coast?

Hanging out there in the wind was one of those single-line rubrics of accomplishment so beloved by adventurers and *The Guinness Book of Records.* I would become "the first person to cross Antarctica on foot." There would be a few asterisks and quid pro quos attached to that description—such as the fact I hadn't done the trek all in one go—but I could live with that.

Somewhere deep in my gut I had the real fear of the adventurer motivating me—the fear that someone else would do it first.

Sir Ranulph Fiennes, described by *Guinness* as "the world's greatest explorer," had succumbed to the lure of the same descriptive phrase in 1992. Fiennes and my onetime Jack Hayward Base hut mate, Dr. Mike Stroud, attempted to man-haul across the entire Antarctic continent, from the Weddell Sea to the Ross, seventeen hundred unsupported miles. They almost died in the attempt, giving up 289 nautical miles from their goal.

An anthem of my university years had been Madness's song "One Step Beyond," the band's cover of a reggae classic by the great Prince Buster, a.k.a. Cecil Bustamente Campbell. A somewhat nutty and insanely catchy ska-beat ditty, "One Step Beyond" had a mood that fit perfectly with my quixotic attempt to complete my South Pole trek. Lead vocalist Suggs's mocking title refrain was perfect marching music. I had briefly adopted it as one of my sledging mantras on Footsteps of Scott. "One step . . ."—*pull*—"beyond!" "One step"—*pull*—"beyond!"

The One Step Beyond Expedition would put me in the record books as the first person to cross Antarctica on foot. But the expedition would also be the first step to honor the commitment I made in Rio to undertake a global mission and a local one, then to come back to the next summit showing what was possible to achieve working with young people. Rio's "think global, act local" slogan led us to develop a mission in Antarctica, then go on to South Africa and develop a local one.

"But I don't want to walk," I said to Pete Malcolm, feeling him out on the feasibility of the feat.

Pete just shook his head. "You want to cross the continent on foot, but you don't want to walk?" He had moved to Australia and was heavy into his yurt period, a time when he built, bought, and promoted the circular Mongolian tent as the most elegant solution to the world's housing needs.

"Let's go into the yurt," I used to say to him. "They can't corner us there."

When I pitched him the One Step Beyond Expedition, Pete didn't understand. No one else did either. It was as though I was talking in riddles. Cross Antarctica on foot without walking? It wasn't until I stumbled across a magazine article about parasailing that I figured out the answer to that riddle.

Wind power.

It was the perfect answer. It was green, renewable, in abundant supply, and it obviated walking.

From my British Antarctic Survey contacts, I recruited a pair of up-for-anything polar veterans named Crispin Day and Geoff Somers. For sponsorship we hooked up with a great company called Spider for parasailing equipment and another business firm, Tandem Computers (since absorbed by Compaq and Hewlett-Packard), under the leadership of Roel Pieper, now a great friend.

The Tandem One Step Beyond Expedition quite literally hooked up with Spider, since the parasail company made a product that hooked onto a harness and pulled one along effortlessly. More properly, the company's ingenious rig was a modular parafoil, "a multi-cell wing type aerial device," according to the original patent. By adding or subtracting elements to the sail, the user could adapt to wind conditions.

Originally, the parafoil was used in tandem with something called a parakart—not to be confused with a parakeet—a small set of wheels for use on the beaches of the world. "Sand surfing" was popular in Australia and other venues with long, windswept beaches. But when it was paired with a snow sledge, the rig was clipped to a parasailor's chest harness, and the polyester bridles were attached to a sail of rip-stop nylon. Two brake lines and two power lines controlled the kite. We would be able to stuff the unused sail portions into a compression pod that we wore on our backs or kept on the sledge, out of the way but ready for instant access.

One of the primary issues with a parasail was not sailing but stopping. The kites rode high enough in the air that they accessed the steady laminar flow of winds that essentially never quit. I had images of being dragged for miles across the Antarctic Plateau, riding the oblivion express. To prevent this from happening, an essential part of our kit was a stake we could drive into the ice to halt our progress.

Geoff and Crispin signed up as my expedition mates. Geoff Somers had undertaken one of the great Antarctic journeys, successfully transversing the continent with dog sleds under the command of my great friend Will Steger. A lanky, angular bloke, he always reminded me a little of Prince Charles.

I used to torment the other member of the expedition, Crispin Day, by quoting bits of Shakespeare's famous Saint Crispin's Day speech from *Henry V*:

> *Old men forget: yet all shall be forgot,*
> *But he'll remember with advantages,*
> *What feats he did that day.*

We trained with wheeled in-line skates on the windblown, seven-mile-long Cefn Sidan Sands beach of Pembrey, Wales.

After a hard day of parasailing, I'd mutter under my breath, just loud enough for Crispin to hear me, "We few, we happy few, we band of brothers . . ." I'm quite sure I was the first person ever in the history of the planet who had the idea to quote the Saint Crispin's Day speech to a guy named Crispin Day. He suffered my boorishness with great good nature.

Crispin was able to master the technique of sledging upwind with the parasail. I could never really get the hang of that. I hoped the steady flow of southerly winds off the Antarctic Plateau would hold for the duration of the trek, and I wouldn't have to be kiting upwind anywhere.

"I'll agree to do this expedition on one condition," I told Geoff and Crispin. "The condition is that I don't want to walk a single step."

"Come one, Rob," Geoff said. "The bloody thing is called One Step Beyond."

"Maybe we should name it the Not One Step Expedition," Crispin said.

I was serious. After Icewalk and Footsteps, I had had enough walking for a lifetime.

The anal-retentive attitude of the American government that I had encountered on Footsteps of Scott had, in the intervening decade since 1986, relaxed somewhat. In early December 1996, Crispin, Geoff, and I flew into Amundsen-Scott Station without getting tangled in too much red tape. No ships this time; no wintering over. And, I insisted over and over, no walking.

In the flurry of preparation for the expedition, I belatedly realized how unprepared I was for what greeted me on my return to the pole, ten years after my first visit.

The South Pole

The geographic South Pole, latitude 90° S, is set not in stone but in ice. Because the Antarctic ice cap drifts to the north, some thirty feet per year, the marker has to be moved every summer to reflect true south.

Adding to the uncertainty is that the planet wobbles a bit (fifty-five feet, more or less), so that the axis of rotation is never totally precise. Which means that the south geographic pole is never truly precise, not in the way that scientific measurement demands. It's a layman's South Pole, a people's point of reference.

Yes, there is indeed a pole at the South Pole, twelve feet high, embedded in the ice pack outside the Amundsen-Scott Station at some point called "the ceremonial South Pole"—that is, a place for photo opportunities. It's a candy-cane barber style of a marker, something of a wry joke, topped with a customized medallion that is redesigned every year by the members of the base who have wintered over.

At the geographic South Pole itself flies the U.S. flag, a "just-in-case" territorial marker that maintains an American claim of suzerainty in all sectors of Antarctica, should Washington ever decide such a claim needs to be made.

The pole sign is there, too:

Geographic South Pole
Roald Amundsen, December 14, 1911
"So we arrived and were able to plant our flag
at the geographical South Pole."
Robert F. Scott, January 17, 1912
"The Pole, yes, but under very different
circumstances from those expected."

Nowadays a signature smell hangs in the air around the South Pole: the perfume of JP-8, the jet fuel cached in drums and burned by the resupply Hercs that fly in and out constantly in the Antarctic summer.

The massive $162 million Amundsen-Scott base, with stations, observatories, and roads, completely surrounds the pole. Research disciplines include glaciology, biology and medicine, geology and geophysics, climate studies, astronomy, and astrophysics. The place does superb science, and I respect them for it. But there's not a solar panel or wind turbine anywhere on the base—this in one of the windiest places on the planet with twenty-four-hour sun for long periods during the year.

My return to the pole with Crispin and Geoff for the One Step Beyond Expedition was the first time I had been there since In the Footsteps of Scott. It was a full decade later. The big American base sprawled around me. This time around, the attitude of the Americans was totally different. We were welcomed with open arms. There were only two or three among the base personnel who had been there a decade earlier. But from each one, and from a few other people I had never met before, I heard a similar sentiment. *We're sorry how you were treated the last time you were here.*

A lot had changed in the interim. Adjacent to the geodesic dome that still housed many of the facilities, the Americans had erected a huge, four-winged, two-story housing unit, the whole modular building designed to be jacked up above the accumulating snow. Ugly and buff-colored but serviceable, it resembles a small suburban apartment building, a little slice of Akron, Ohio, say, dropped down on the South Pole.

Since I had last been there, the gates of Antarctic tourism had opened. In the Footsteps of Scott had indeed turned out to be a harbinger, an icebreaker. Giles Kershaw started what amounted to the world's first Antarctic airline.

Giles ferried mountain climbers, expeditions, and adventure travelers to the continent. He flew climbers to Antarctica's highest peak, Mount Vinson. He maintained a private seasonal base at Patriot Hills, in the Antarctic interior. He always did it right, carefully protecting the fragile polar environment, successfully seeking to educate their expedition members more than providing them with cheap tourist thrills.

As a pioneer of private enterprise in Antarctica, Giles established a loose, quid pro quo relationship with the scientific community. Not a few times, he rescued marooned researchers, including a memorable occasion when he whisked a pair of South African scientists off an iceberg where they had been stranded for eight days.

When Giles died, his wife, Annie (my future CEO), inherited the first Antarctica expedition company, Adventure Network International, the legendary ANI.

The former Anne Campbell was a Britannia Airways flight attendant with an engineering degree from Glasgow University. Her husband died in Antarctica in 1990, killed in a crash while flying a photo-support gyrocopter for an expedition in Jones Sound. He was two hundred feet in the air when a wind gust took him down.

Annie directed that Giles be buried beneath a stone cairn, on a rock ledge near where he crashed, below a mountain that would later be named for him.

"That kind of burial could probably never happen again," Annie told a reporter later. "Today, there would be a big debate over it. Is his body history, or garbage?"

In Giles's case, I myself would vote for history. He was the greatest pilot Antarctica had ever seen.

An amazing woman, our Annie, warmhearted and hard-nosed in equal parts. As Giles would have no doubt wanted her to, Annie forged on after his death. Working from Punta Arenas, Chile, she chartered public expeditions to Antarctica, opening the continent to the wider public for the first time.

It required a delicate balancing act. How do you expose people to the pristine wilderness of Antarctica without rendering it less pristine? Annie was a pioneer. She believed that the benefits of getting people to support the preservation of Antarctica far outweighed the risks associated with bringing them down to see it.

So in the decade since I had last been in Antarctica, government-funded scientists had slowly begun to share the continent with scores of private visitors. I noticed that the garbage around Amundsen-Scott had been tidied up. Perhaps exposure to the public was not so bad after all.

In a single decade (1986–96), Annie had opened the doors for private expeditions. No longer did trekkers need a ship, they could make journeys supported by aircraft. As we prepared to embark on One Step

Beyond, I recalled my aborted arrival at the base ten years earlier. I remembered something Mark Fish had told me. "You never really got to the South Pole," he said, meaning my achieving the pole was utterly changed by circumstance. Within minutes of reaching it the first time, I had learned that the *Southern Quest* had sunk.

During my second visit in 1996, everything around me appeared new. I walked out from the station dome to the South Pole marker. The last time I had been there, everything had been a blur. It was as though I had walked up to the pole in 1986 and someone had immediately whacked me from behind with a two-by-four. I never enjoyed the success of getting to the Pole. I was utterly despondent.

When Captain Scott finally made the pole, he, too, experienced a devastating sense of disappointment. Amundsen had stolen a march on him, heading up the triumphant first expedition to reach the South Pole. After his physically gruelling journey, Scott had attained his goal, only to see Amundsen's abandoned black-silk tent already there.

"It was a little masterpiece of a tent," wrote Amundsen of this marker of Scott's disappointment, "made of thin silk, which, when folded together, would easily have gone into a fair-sized pocket, and weighed hardly a kilogram."

Seeing that black spot amid the death-white emptiness of 90 degrees south must have crippled Scott's spirit. Perhaps *spoiled* is too strong a word, but the fact that he had lost his primacy certainly affected any sense of achievement Scott might have felt. "The pole, yes, but under very different circumstances from those expected."

I thought of the little black tent, and of the *Southern Quest* lying at the bottom of unquiet seas, as I strode out alone from the base during my second visit in 1996. I looked pretty ridiculous, since I wore the same hat that Kurt Russell wore in the movie *The Thing*, an Australian bush hat on acid. I carried Teddy in my pocket, the second time the little bear had been to the South Pole.

Walking north from true south, I proceeded a hundred yards to the same point on the ice cap where the South Pole had been located ten years before, when I had first trekked to it. Because of the steady movement of the ice cap, each step was like walking back in time, a second coming to a spot I had been once before.

T. S. Eliot's line came back to me: "And the end of all our exploring/ Will be to arrive where we started/And know the place for the first time."

The only effective method to treat post-traumatic stress seems to be to reexperience the event in some way, either through guided memory, EMDR, hypnosis, or talk therapy. As I stood on the ice where I had stood back then, I again let the pain of loss, guilt, and frustration over the *Southern Quest* overwhelm me. I tried consciously to go deep into it, again heeding Frosts's line "The best way out is always through."

Then I did cry, once again. Thanks, Roger; thanks, Gareth; thanks, Pete and Will and Rebecca and Lavinia and Samantha—thank you for this moment that I never experienced before.

Then I let it go. I ventured out farther along the same ice-cap time line, hundreds of yards, a half mile, farther and farther until I arrived at the spot that Scott and Amundsen had reached in the summer of 1911–12. Here was where my heroes had stood. Somewhere in the ice down below I could have excavated their footsteps, discovered the stubs of their victory cigars. I could have exhumed the black tent Amundsen had left behind for Scott, now almost a hundred years snowed over.

As a schoolboy in 1966, I decided to follow Scott to the pole. I had stuffed thirty years of Scott, Shackleton, and Amundsen into my brain. I celebrated them all at that moment, standing above the migrated ice of their historical South Pole.

Scott, you got here, you made it, but you did not make it back.

Amundsen, you were the most brilliant of them all.

Shackleton, you never made either pole, but you persevered through hardship I can barely imagine.

I brimmed with enthusiasm for the trek ahead of me. I felt as though I was going back to the future. As Ross had used sails to power his way through the ice pack and see the barrier for the first time, I would use sails to complete my crossing of the continent.

But as I had discovered many times in the course of my expedition life, the path of true adventure never did run smooth. Even though we would have stretches of easy sailing on One Step Beyond, at the end it would all come to smash.

Kiting

———

Because I liked to make my life complicated, and because I had told the world leaders at Rio that I was going to do it, I had to figure a youth element into One Step Beyond.

Perhaps I should have learned my lesson on Icewalk, when the serene idealism of the students burst the bounds of their mandate to sit still and passively learn about the polar regions. Instead, their passion and fire required them to actually do something concrete to address the situation. I admired them for their determination. As with all learning environments, the students on Icewalk taught the teacher a thing or two.

After Rio I organized another international group of students for an Antarctica expedition, thirty-five young people from twenty-five nations. This was a first tentative step in keeping my resolve to introduce the decision makers of tomorrow to the 2041 concept. The students I enlisted on Mission Antarctica would be the blade runners of a whole new generation of polar advocacy.

Or so I hoped. The expedition was organized under the aegis of UNESCO. As a UNESCO envoy, I thought the responsible thing would be to get some input in order to settle upon the global mission that I had quite cavalierly assigned myself in front of the world leaders at the Earth Summit. Who better to help me figure out what our global mission should be than the students on the UNESCO trip?

I made a concerted effort to bring together students from warring nations. Because we were taking them to a place we should never fight over, we recruited a student from Russia and one from Chechnya, a Catholic and a Protestant from Northern Ireland, a Bosnian, a Serb, and a Croat.

The plan was simple. At the same time the UNESCO students were steaming from South America to the Antarctic Peninsula, I would be nearing the coast myself, kiting south with Geoff and Crispin. If all went well, our paths would intersect, and after a short airplane ride I would drop in among the students like a superhero.

Man makes plans and God laughs—isn't that the saying? Antarctica has been giving God belly laughs in this respect ever since man first discovered it. *Nothing* goes according to plan in the polar regions, south or north. Swan's Law: The higher the latitude, the more things go awry.

The first step of One Step Beyond came off without a stumble. The sledges were heavier than we had intended: 330 pounds, over a hundred pounds more than anticipated. God chuckled. We would be supported by three resupply flights over the course of the six-week trip—and still our sledges broke the scale. As with Footsteps, we were again attempting a march unassisted by dogs or engines.

Day one was a breeze, as they say in parasailing. I couldn't quite believe it. Why hadn't I always traveled by wind? The Spider rig transferred the pull of the kite to the harness without chafing or pressure points. The buildings of Amundsen-Scott dwindled behind us. We were left alone in the glabrous immensity of the plateau.

The Antarctic Plateau and parasailing were a peerless match. There ought to be signs erected: PLEASE PARASAIL HERE. We would stop and stake our rigs every hour, change our sail configurations, adjust our relative speeds in order to stay in a group. It felt great.

I was not quite as adept as Crispin and Geoff. Once again, as with Roger and Misha and so many others, I partnered with experts who knew quite a bit more than I did about the task at hand.

Luckily, the kites worked on a "dead man's" system. A morose term for a fiendishly ingenious method preventing the sail from kiting away from a downed rider. As soon as the handles that controlled the kite were released (as when I dumped myself unceremoniously on the ground), the canopies above me collapsed. That, along with the practice of staking the kite, prevented me from being dragged off by a runaway parasail. We also all carried a huge knife sheathed at our shoulder, so if all was lost, we could sever the leads.

So far, so smooth. We made ten and a half miles the first day. It was more exciting to go faster, and very tempting. But slow and steady did it. We eventually settled in at around five miles an hour.

I often dwell on the attraction that certain forms of locomotion have for young people. My son, Barney, is a great skateboarder, snowboarder, and downhill mountain biker. There is something simpatico about benign natural forces such as gravity and wave action. To anyone under twenty, they must appear to be like gifts from God. No mining or drilling necessary. No asking for the car keys. Plenty of young people are attracted by the internal combustion engine, but an equal number seem to migrate toward more natural forms of transport. It's as if they are born green.

Barney will love this, I thought on that first, exhilarating day of sailing. I tucked away that insight for the future.

It all came tumbling down the next day. No wind.

"No wind?" I raged. "No wind? This is the Antarctic Plateau. There's *always* wind."

But on December 12, 1996, there was not a whiff of a breeze to be had. I recalled morosely that Scott had affixed a sail to his sledge on his doomed return trip to the pole, but the winds had failed him.

We were near base, still within a short day's march to Amundsen-Scott. We could turn back, couldn't we?

The Spider parasail harness had been specifically reengineered to double as a man-hauling rig. In the course of our preparations for One Step Beyond, I had either missed this detail or suppressed it out of a sense of denial.

"We're not going back, and we're not waiting for wind," Geoff Somers said, and he hitched his sledge to his harness and set off.

Suddenly I was sucked back into the same sense of despair that I had on Icewalk. I simply could not face the mind-numbing drudgery of man-hauling.

One step . . . beyond!

I put my head down and put my shoulder into it. I didn't cry this time, as I had on Icewalk. The plateau stretched out endlessly before me. I was in the movie *Groundhog Day,* condemned to repeat the same excruciating trek over and over. I was feeling very dark indeed.

As it turned out, that second day was simply a test that Antarctica had thrown in our path. A way to keep us honest, a way to indicate that things can't be too easy at the bottom of the world. From day three on, we sailed. Literally, and with an increasing sense of elation. We did 31 miles that day. We had a few days of no wind, including Christmas Day, 1996. But on the next day we did 62 miles, and on January 4, 1997, an incredible 102 miles.

Manipulating the kite, changing the components of the sail a few times daily to adjust for changing conditions, Geoff, Crispin, and I found we could make excellent time and still stick together—a band of brothers. More surprising, we discovered we could use the parasail crossing fields of sastrugi, a skill that even Geoff Somers had thought impossible.

Churning steadily forward at five, six, or even ten or more miles per hour, I leaned back in that harness and grinned. It was not quite effortless but nearly so, and so much less laborious than walking that I swore I would take up the parasail for all my transportation needs. Our caloric requirements were so reduced from lack of exertion that we found we could not consume all our rations.

But still Antarctica wore on me. While the physical toil was removed or lessened, the effect of the psychological effort asserted itself. I found myself snapping in and out of time, experiencing the kind of disorientation that long-distance truckers feel, "white-line fever"—in my case, white-landscape fever.

I had moments of not knowing where I was. Was this Icewalk? In the Footsteps of Scott? The Arctic? The plateau? The Great Ice Barrier?

It was decision time. There were two ways of looking at the choice I had to make. I was three hundred miles from being "the first person ever in history to cross all of Antarctica on foot, et cetera." We had already accomplished seven hundred miles of a thousand-mile trek. We could do this. I knew we could.

The other side of the equation was the thirty-five young people I was scheduled to meet, the UNESCO group. I couldn't let them down, could I? We had been considerably delayed by bad weather getting to the pole. Now it looked as though I would be late for the rendezvous.

My mind played tricks on me. Was I really concerned about the UNESCO kids, or was I just looking for a way out? The "get me out of here" element could not be dismissed. Every day I awoke to a sense of dread. *If I see one more mile of Antarctica,* I found myself thinking, *I am going to blow up.*

The blank-faced god of Antarctica seemed to mock me. *Why do you want to leave? Which is it, Robert? The UNESCO expedition or the get-me-out screaming heebie-jeebies?*

It was not the easiest decision I ever made. I truly missed my wife and son. I had put so much into One Step Beyond. I knew someone else would whip in and cross the continent on foot. I had always been motivated at least in part by fear. Fear of failure, fear of seeing the other bloke accomplish what I had tried and failed to do.

I've already done it, I thought. What does it really mean? It was my modest, watered-down "live donkey or dead lion" moment, reminiscent of Shackleton turning back ninety-seven miles from the pole. But why did I even have these people in my brain? Long-dead Heroic Age explorers who no one even remembered—would they dog me all my life?

No. This was going to be one moment when I would just say, for once, *What do I want to do?*

What I was going to do was fail for the first time in my polar career. I wasn't going to get there. Dealing with failure was something I never experienced. Here it was, staring me in the face.

In ten years, or perhaps less, the UNESCO kids were probably going to forget they ever went to Antarctica. The meaning of their trip would vanish, while I was dropping out of something that meant history—a small, tiny sliver of history but history nonetheless.

It is difficult for somebody not obsessed with expeditions to grasp what something like "the first person to cross Antarctica on foot" might mean. Looking at it on the page, it seems slight, presumptuous, a ridiculously thin claim to fame. But ever since I was a child, I had been obsessed with making my life *mean* something. A vain hope, I know, and an egotistical illusion to boot. But maybe precisely because I was aware of the essential meaninglessness of the universe did I cling to these tiny shreds of accomplishment.

There were leadership lessons lodged in there somewhere. Number one: Never give yourself an easy way out. The UNESCO expedition was a convenient excuse. Second, don't try to dress up your decisions.

Be true to you. If you have to quit, accept that in yourself. That is what you do.

I needed to get out of there. "I have to go meet the kids," I told Geoff and Crispin. We had been becalmed for the day by light head-winds.

Geoff looked at me as though I was mad. "Really?" he asked.

Crispin seemed to understand better. "You don't have anything to prove," he said.

It happened very quickly. I contacted Anne Kershaw via Argos beacon, and she sorted out an airplane. The two boyos carried on parasailing. Seven days later, they reached the coast of the Weddell Sea, having completed the expedition without me.

Bellingshausen

Antarctica has a garbage problem.

At all of the year-round Antarctic research stations (some thirty-five exist) are equipment dumps, trash depots, storage tanks of human waste. With little or no decay bacteria present in the Antarctic environment, nothing organic breaks down. It's like Las Vegas: What happens in Antarctica stays in Antarctica, pretty much forever. The choice is to haul it out or to leave it.

Leaving it is cheaper, so for the hundred-plus years of human habitation on the continent, the trash piles have been slowly building and the middens have become increasingly foul. To witness the modern ecological conundrum in microcosm, to see the environmental problems of industrial society physically laid out before your eyes, there is no better place to go than Antarctica.

The same is true for both polar regions. Icewalk's student contingent reacted with thoroughgoing disgust at the mounds of garbage surrounding the Eureka base, the result of years of accumulation. With the precious naïveté of the young, they immediately conducted a letter-writing campaign, challenging the Canadian government in Ottawa to clean up the mess. Incredibly, it worked, and a few tons of trash were hauled out of Eureka and transferred to landfills.

The Icewalk students didn't stop at writing letters. All over the northern territories of Canada, dumps of rusting, leaking fifty-gallon fuel barrels represent a legacy of years of bush-pilot flights. During their time at Eureka, the students journeyed forth and actually got their hands dirty, wrestling the barrels from dump to plane to recycling center. They barely scratched the surface of the problem, but they did indeed scratch it.

When I connected with my UNESCO students after bailing out of One Step Beyond, I found them terrifically excited to be journeying to Antarctica. I met the group in Punta Arenas, Chile, at the southern tip of South America. Our departure was delayed by problems with the ship—so it turned out I could have finished One Step Beyond after all. One of the Immortals, John Tolson, who had documented Footsteps, just happened to be there at the same time and had helped sort out the ship for the expedition—just another example of what goes around coming around.

Our voyage across the fearsome Drake Passage—notorious as the worst seas on earth—was interesting for the friendships it spawned. An odd thing happened: Supposed opposites attracted. The Bosnian student chose to bunk with the Croat. The Palestinian made friends with the Israeli. The "enemies" tended to develop closer friendships with each other than with any of the "neutral" students on the expedition.

But their first encounter with Antarctica stunned them all to silence. We steamed into Collins Harbor on King George Island's Maxwell Bay. The thirty-five students from twenty-five countries gathered at the bow of the ship, like a row of toy soldiers bundled in down jackets against the cold.

Stretched before us was . . . garbage. Acres of it, tons of it, mostly machine parts and industrial trash, spread around the Russian base at Bellingshausen. The accumulations of cold war–era rubbish at the station represented a sobering first look at the continent for the UNESCO students. The dump covered nine acres and represented a slap-in-the-face introduction to Antarctica.

It proved to be an indelible image throughout the trip. Yes, yes, the kids were exhilarated at the stark raving beauty of the landscapes they were seeing, but they kept coming back to the issue of garbage. They reminded me of someone worrying a bad tooth. Wow, a humpback whale breaching—but can you believe the mess we saw back there at the Russian station?

"Smellinghausen," the students had nicknamed Bellingshausen.

I thought a good global mission might be to adopt a specific species of penguin, maybe do a little informational booklet about it, perhaps a short video. But the mix of idealism and recklessness, often remarked upon in those under the age of thirty, moved the expedition students to suggest a slightly more ambitious agenda.

If McMurdo, with its big New Zealand and American bases located next door to each other, can be said to be Antarctica's New York City, then Bellingshausen, on the other side of the continent off the tip of the Antarctic Peninsula, is its Washington, D.C. More than a dozen international bases in total and a half dozen year-round Antarctica stations are located in the South Shetland Islands, near Bellingshausen on King George Island. The South Shetlands enjoy the mildest climate on Antarctica (average mean temperature in winter, a balmy 23 degrees F). The Russians have nicknamed their Bellingshausen base the *Kurort*, "the resort."

Bellingshausen is a natural first stop for expeditions and tourism cruises departing from Chile or Argentina at the southern tip of South America. The voyage between Ushuaia and Bellingshausen is the shortest to Antarctica from any other continent, even if it does cross the fearsome, storm-tossed Drake Passage.

Featured in the vast jumble of rubbish were assorted engine blocks, sheet metal, lengths of pipe, discarded water heaters, ship and aircraft propellers, radiators, several hundred fuel barrels, empty steel-cable spools, windlasses and broken winches, drums of discarded mechanical lubricants, commercial fishing tackle, four truck chassis, fuel hoses, prefabricated-housing panels, a number of axles and drive trains, and a half-crushed portable outhouse.

The Soviet government established Bellingshausen Station in 1968. Like most official Soviet and American actions during the cold war period, the founding of a permanent base in Antarctica represented a move in the international chess match of the type that was once called the Great Game or, by the Russians, the Tournament of Shadows.

America had created its Amundsen-Scott South Pole Station a decade earlier. The Soviets responded by establishing a base at another pole: Antarctica's pole of inaccessibility, the most remote point on the continent. During the cold war there was a lot of schoolboy puffing up on both sides. The United States was increasing its presence at McMurdo. Every international action demanded an equal and opposite reaction from the opposing side. So the Soviets landed on the Fildes

Peninsula on King George Island, which had the virtue of being relatively ice-free and therefore accessible year-round.

The junk immediately began to pile up. When the government erected a fuel-tank farm at the station, it became a busy refueling stop for Russia's Antarctic fishing fleet, increasing the deposits of rubbish in the dump.

Then, in 1989, the Berlin Wall fell and the political landscape of the world changed. Support for Bellingshausen Station tapered off. Prospects for government action on cleaning up the trash, never very strong, disappeared to nil.

There is a Zen koan, or instructive fable, that tells of a student asking his teacher, "Master, what is the meaning of the universe?" The master replies gruffly, "Clean your rice bowl!"

The meaning I always took away from this is not to get tripped up by the big questions but to remember to focus on the mundane task at hand. Nothing could be less romantic, and more arduous, than cleaning up a mess. But nothing could be more meaningful.

In this case, once again, the students taught the teacher. "How about if we just clean this up?" one of the expedition students asked me, gesturing out the window to the equipment dump. We were in the salon of the ship that had taken us to Bellingshausen. "Make that our global mission?"

Murmurs of agreement all around. "It'd be great! It would totally highlight the problem with garbage around the globe!"

"The chinstrap penguin," I muttered. "I was thinking more along the lines of doing something with the chinstrap penguin . . ."

My words were lost in the excited chatter of thirty-five twenty-somethings all talking at once.

The *2041*

Once again, I needed a boat.

The last time I had felt that urge, when I found out that the only way to accomplish In the Footsteps of Scott would be to ship ourselves down there, Pete Malcolm came up with the *Southern Quest*. This time around, I was looking for something smaller, sleeker, more environmentally sound. I never considered buying a barge or a freighter to haul the trash out of Bellingshausen.

What I needed more than that was money. In fact, I might wind up needing a barge just to haul all the currency it was going to require to clean up Bellingshausen. The easiest way for the 2041 organization to raise funds, I decided, would be to utilize a flagship, a floating fundraising spectacular, a harborside shout-out, our very own 2041 yacht.

Okay, that word, *yacht*. Everywhere in the world apart from America, *yacht* merely means one thing: a watercraft of more than thirty-five feet in length. In America, it means a billionaire sunning himself while servants serve mai tais to girls gone wild in the hot tub. It means riches beyond wealth, Bill Gates or Sergey Brin wealth.

I want to suggest that all Americans reading this should please calm down. Yes, I was looking for a yacht. But no, that didn't mean I was moving up a few tax brackets. I still remained a borderline-bankrupt expedition leader who plowed every shekel of what he made back into his advocacy organization.

But that advocacy organization was in need of serious coin if it was going to embark on a project as ruinously expensive as cleaning up Bellingshausen Station.

In 1998, the 2041 effort found itself a flagship. Neelie Kroes, the commissioner of competition for the European Union, heard of the Bellingshausen project. What do you need? she asked.

"I need a sailboat," I told her, initially avoiding that red-flag word *yacht*, even though Neelie was Dutch and thus perfectly conversant in maritime nomenclature. "I need a boat for fund-raising and to ferry young people down to Antarctica."

Neelie is a fantastic woman, one of the real determined doers in the world. Her home country, the Netherlands, is often seen as overwhelmingly liberal and politically correct, but I noticed that it did not have many women in the upper echelons of business or government. It was traditional in that way. Neelie was the exception. She had fabulous contacts and put me in a roomful of the most senior Dutch businessmen she could find.

I made my pitch. "Look, in this country, everyone has a car in the garage," I said. "If I lived in this country I would have a boat in the backyard, too, because you guys are going to swim first." Just as they and their below-sea-level countrymen considered the ability to travel by water a personal necessity, so too did 2041 consider it an organizational necessity.

And just like that, it was done. The group of Dutch businessmen (from some of the biggest companies in the Netherlands—Phillips, Unilever, Robo Bank, and KPMG) found the *2041,* a beautiful sixty-seven-foot racing yacht. I remember going with my friends Jan and Nicolette to see her for the first time where she was moored dockside in Amsterdam. An outmoded nautical term bubbled up through my unconscious. This boat was yar—meaning sleek, well built, and seaworthy.

John Howard (Kittredge) and Katharine Hepburn (Tracy) in *The Philadelphia Story* use the term in talking about the yacht the *True Love:*

> TRACY: *My, she was yar.*
> KITTREDGE: *"Yar"? What's that mean?*
> TRACY: *It means, uh— Oh, what does it mean?*
> *Easy to handle, quick to the helm. Fast, bright.*
> *Everything a boat should be.*

She was yar, all right. A steel-framed Bermudan cutter designed by David Thomas and built by Devonport Management, in a Plymouth shipyard in 1990, for the Chay Blyth Global Challenge round-the-world yacht race. Below deck, a Perkins-Sabre M-185C turbocharged diesel engine, six-cylinder inboard, and berths for fourteen in six cabins. Built for one of the most rigorous yacht races in the world, a thirty-thousand-mile against-the-prevailing-winds endurance trial, she had a cruising speed of 7.5 knots.

She was yar, but we wanted to make her green, too. We went to work refitting our new flagship with wind turbines for electrical power, alcohol stoves for heat and cooking, and, eventually, recycled polyethylene terephthalate (PET) plastic for sails.

We didn't name her the *True Love*. We rechristened her the *2041*, and the yacht immediately set sail for Antarctica with a student crew of volunteers, embarking on a circumnavigation of the globe that she would not complete until 2008. We were sent off at our launch by the heir apparent to the Dutch throne, Prince Willem-Alexander.

2041, the yacht, helped immensely with 2041, the organization. The graceful sailboat, parked dockside at harbors around the world, acted as a floating billboard for the cause. She also acted as a classroom in the cold for Internet presentations. She facilitated the Bellingshausen cleanup, ferrying volunteers to Antarctica, and eventually helped to raise millions of dollars for the effort. But she would also accomplish one of the most surreal—and so far unrivaled—journeys of any boat in history.

Mission Antarctica

W hy are you doing this?" the official demanded. I was in Moscow, seeking the Russian government's permission to help address the mess at Bellingshausen Station. It was the aftermath of the cold war. The dismantling of the Soviet Empire put the whole region into disarray. Yeltsin's tanks had only recently been in the streets.

"I want to join the effort to clean up Antarctica," I said. I had been met so far with only suspicion and resentment. Why was I here? What were my real motives? Was I out to embarrass the Russian government?

I knew the conditions at Bellingshausen. Financial backing for the superb science being conducted at the station had dried up. The staff there dwindled and those that remained were relatively impoverished. The base had become well known in polar circles as a place to trade supplies for Soviet Army souvenirs.

In Moscow, when I visited, I witnessed World War II veterans selling their military medals in the street. The choice confronting the government was to clean up the trash at Bellingshausen or continue, even on a bare-bones budget, the great science program there. The government naturally chose science over trash removal. Now here I was, offering them a free ticket to a clean beach. All I needed was permission.

It was finally given, albeit grudgingly. *Keep your mouth shut about it,* was the essence of the message I got from Russian officialdom. The whole country had been humbled. It was as though its collective pride was hurt by having to accept help. The Russians would only agree to me helping because I was a private citizen.

Misha Malakhov, the Immortal of all Immortals, helped get his countrymen on board. He had just been made a Hero of Russia. That's not a descriptive tag, it's a real official honor, elevating the recipient into an elite category. People stand up when a Hero of Russia walks into the room. Misha was no longer required to pay utility charges, hotel bills, or transportation fees (of course, Misha being Misha, he accepted none of these, and went right on paying his bus fares and the rest like an ordinary citizen).

The cleanup of Bellingshausen Station became, behind the pole walks, the most physically gruelling task I had ever embarked upon. It developed into an expedition in and of itself. "Mission Antarctica" we called it. Over the course of four years, from 1998 until completion of the effort in 2002, in a joint operation with the Russian government, I raised funds (finally totaling six million dollars), cajoled officials, tugged on scientific sleeves, and got myself down and dirty removing and recycling the sprawling rubbish heaps at Bellingshausen.

But even ahead of the pole walks, it is the accomplishment in my life of which I am most proud. Every Antarctic summer (which means every Northern Hemisphere winter) for five years I worked to load trash into transportable bins, barges, or Dumpsters. Mission Antarctica drew participants from all walks of life: schoolteachers, engineers, and professional sailors. Volunteer welders attacked larger, more unmanageable pieces, cutting them down to size. The Russian government also provided workers, as well as logistical help and quarters for the effort.

And the Immortals came winging back into my life: Not only Misha, who helped liaison with Russians at Bellingshausen, but John Tolson, the documentarian on In the Footsteps of Scott, and Peter Malcolm, now living in Australia and proving an immense help with the naval side of things. A superb organizer named Adrian Evans administered the effort back in Britain. Adrian's brother Gary and his team scrubbed the beach clean.

Going in, I had no idea what an appalling effort it would prove to be. Leadership at times requires a brand of blind optimism. Knowing what a specific job really entails in all its gritty detail would prevent a lot of tasks from getting done. Some of history's great accomplishments have their genesis in ignorance.

I have my doubts whether, with a complete knowledge of foresight, the Normandy landing would have taken place. Would it have just seemed too horrible to embark upon? The Apollo space program?

Napoléon's invasion of Russia? An Ironman competition? The slings and arrows of the average life? On a totally different scale, I know for certain that if I had known what I was getting into, I would have never walked to the poles. Ignorance, at times, is bliss. It's also a precondition for embarking on endeavors of a certain scale.

The weather in the "resort" of Bellingshausen repeatedly betrayed us. Sleet storms and blizzards lashed at the barges we used to ferry the scrap out to our ship, the Danish vessel *Anne Boye*. When officials from nearby King George Island bases (Uruguay's Artigas, Chile's Frei Montalva, and Poland's Arctowski) suggested that they, too, had random scrap dumps to remove, we responded wearily that we would take that trash, also. Sure, of course, pile it on!

Here is a good message for fund-raisers everywhere: Keep in touch with your backers. And: Don't immediately ask a wealthy person for money. You must understand that anyone with a healthy bank account fields several requests for support every day.

Ernesto Bertarelli is the former CEO of Serono, the Swiss biotech company, which he inherited from his father and built into an enormously successful enterprise. I got to know him over the years, primarily through his interest in yacht racing. (In 2003 and again in 2007, Bertarelli would win yachting's grand prize, the America's Cup, for Switzerland, an all the more incredible achievement since the country does not have a coast.) Throughout our association, I had never asked Bertarelli for a contribution. But at the end of the cleanup effort at Bellingshausen Station, I had run into a brick wall. The increased fundraising capability of the *2041* had helped, but not enough.

My primary backers for the effort were American banking and investment firms. Then the 9/11 attacks on the World Trade Center and the Pentagon took place, and the focus of those companies was understandably elsewhere.

The terrorist attacks cast a pall over the whole world. I felt helpless. But I reacted along the lines of a lot of other people—with anger. I simply refused to let the Bin Ladens of the universe dictate my choices. I bulled ahead.

The *Anne Boye* had already set sail toward Bellingshausen. I committed to a last cargo voyage to remove the collected trash out of Antarctica, even though I did not have the money to pay for the ship.

I telephoned Ernesto. "The ship is already on the open seas," I told him. "The bow wave is headed south." Mission Antarctica had floated the cargo ship on a tide of red ink. He came through with a check that put us over the top.

In all, Mission Antarctica took fifteen hundred tons of scrap off the beach at Collins Harbor. Our team carted the tons of rubbish by ship to Uruguay, where the metal and everything else was recycled. Somewhere in the Uruguayan capital of Montevideo, buildings now exist reinforced with steel rebar fabricated from recycled Antarctica scrap.

I continually encounter strong parallels between Antarctica and the wider world. We are everywhere drowning in our own waste, whether it comes from an SUV's exhaust pipe or a sewage-treatment plant. The situation is the same all over the globe. As usual, the picture is just a little clearer in Antarctica.

At the end of the season in 2001, I gazed out upon a swept-clean, pristine section of black volcanic sand where once a towering pile of refuse had stood. We had our own *March of the Penguins*: A flock of the comical birds (and, yes, a few of them were chinstraps) tentatively explored the beach, access to the sea restored to them at Bellingshausen for the first time in decades.

Cape Town (III)

Who were the geniuses who decided to hold the 2002 Earth Summit in Johannesburg, South Africa? The site was a hundred miles from any ocean, in the bushland of the high veldt. Couldn't they have done it somewhere with a harbor?

My global mission was successfully completed in time for Rio +10, as the Johannesburg Earth Summit came to be known. The word *sustainable* had crept to prominence in the bureaucratic lexicon, since the gathering was officially called the World Summit on Sustainable Development. It was March 2002, three months before I was to report back to the U.N. conference on the global and local missions I had been charged with ten years before.

What could I do as a local mission? How could I employ the *2041* in landlocked Johannesburg? The sailboat had become a prime PR asset for environmental change. A lot of good that would do in Cape Town Harbor, when the conference was 870 miles up-country.

The relevance check is a fundamental part of sustainable leadership. Periodically, you've got to ask yourself if what you are doing is relevant, if your stated goals make any sense anymore. If you don't do this, you run the risk of circumstance changing and your leadership efforts getting left behind as outmoded or irrelevant.

What was my leadership message on my local mission to the people around Johannesburg? I could just imagine myself in the dusty townships of South Africa, where poverty is endemic and AIDS was on the rampage, sprinting out, arms raised and smile pasted on face, to address audiences on . . . the importance of cleaning up Antarctica.

I would have been greeted by the sound of crickets. I could not transport my 2041 passion to the local mission. It was a square peg in a round hole. It would not fit. I would cease to be relevant, and therefore I would cease to be a leader.

What to do? Think global, act local. What was my local mission?

Shaun Johnson, a groundbreaking antiapartheid newspaper editor in South Africa, now one of the best-known men in the country, showed me the way. "It's got to be AIDS," he said. "That's got to be the local mission. No question."

He was right, of course. The problem of AIDS throughout Africa staggered the imagination. Seven million people were infected with HIV in South Africa alone, thirty million throughout the continent. But what could I do? I was a sailor and polar traveler, with a sailboat that was all about protecting the environment. What could I do with a yacht?

If the mountain will not come to Muhammad, then Muhammad will go to the mountain. In conversation with team member Angus Buchanan, I developed an outlandish plan. I would take the show on the road.

Angus, an excellent sailor and deckhand on the *2041,* was the voice of reason. I met Angus on the yacht *Pellagic,* captained by the finest of all Antarctic yachtsmen, Skip Novak.

"I've got to go there," I told Angus, wrestling with the idea of how to stay relevant at the Johannesburg Earth Summit.

"Screw it, Rob," Angus said. "Do you really have to? Just go there in person. What are you going to do, truck the damn boat up there?"

Lightbulb.

Partnering with a local activist organization called loveLife (that's how they like to spell it), I planned to dismantle the *2041*'s keel and masts, hoist the boat onto a custom-built yacht trailer, and take it on an overland journey through the parched heart of South Africa, promoting AIDS awareness.

It would be the longest overland ship journey in history. Outlandish, yes, but as it worked out, a surreal stroke of activism that reached 500,000 young people throughout the country.

Under the inspired leadership of David Harrison, the *2041* began its land voyage at Cape Town, at the same harbor where I drove a cab during my student days in the seventies, under the looming flanks of Table Mountain, from which I spied the small red-orange dot of the *Southern Quest* as she arrived for the Footsteps of Scott's voyage to

Antarctica. I recalled the doomed ship in dry dock there, getting refitted, looking like a toy next to the huge supertankers. I felt as though I was living in a continually recycling past.

As if to emphasize the point, two of my Immortals reentered the stage: Seton Bailey, who had been a valued crew member and filmmaker on the *Southern Quest,* and Noelene Kotschan, a.k.a. "No Nonsense Noelene," without whom we would have never gotten the overland voyage of the 2041 launched.

With Seton handling publicity and Noelene on logistics, we surmounted the insurmountable problem of how to haul a fifty-ton boat up hill and over dale. We came up with a special eight-wheeled rig, sixty-seven-feet long, a monster of the highway.

I felt like German filmmaker Werner Herzog in his epic movie of Amazon boat-hauling lunacy, *Fitzcarraldo.* We hitched up the yacht trailer and headed into the interior of South Africa, eventually logging eight thousand miles, the longest overland journey of any boat in history. *Guinness,* take note.

Our arrival in the inland towns provoked storms of interest. We publicized our initiative as "Ice Station," an inviting description in the South African heat. Most of the audiences we spoke in front of had never before seen a boat firsthand, never seen the sea, never seen a photograph of an iceberg or of penguins. It reminded me of when NASA took an Apollo space capsule on a road trip around America. Citizens were gape-jawed. The yacht trailer had a large sign affixed to its front bumper that read ABNORMAL in large red letters—a reference to the oversized load, of course, but I couldn't helping thinking the word could apply equally well to my own sensibility.

I traveled with a student crew, most of whom had just visited Antarctica. They emerged from the yacht into the sweltering heat of the South African autumn dressed in full polar kit, parkas, boots, balaclavas, sunglasses, and mittens. They looked like space aliens. The message: Just as we had to protect ourselves in Antarctica from the cold and wind, you must protect yourselves when having sex.

We were a smash hit. We never had an audience of less than five hundred, and at a few stops we had up to twenty thousand. Through

Afrikaner country, Piketberg, Vanrhynsdorp, Upington, and Kimberley. To the north, to Pretoria, Pietersburg, and Phalaborwa. Down the coast to Durban, East London, and Port Elizabeth. Inland again to Gariep Dam, Welkom, and, finally, to Johannesburg. Forty stops, forty AIDS-awareness presentations, hundreds of thousands of minds reached with a vision, I hoped, that was indelible.

Many people quote that Grateful Dead lyric, "What a long, strange trip it's been." But standing in ninety-degree heat in Warmbaths, South Africa, watching a South African child who had never touched ice before pet a large chunk of it imported from an Antarctic glacier, witnessing a cold-weather dog-and-pony show featuring an international group of down-jacket-clad students delivering a public-health message to a crowd of five thousand teenagers—well, I believed I earned the right to invoke that line.

As a new father, I had a personal reason for reaching out to the next generation. I felt a real sense of optimism all around, since I was heading into a new marriage. My fiancée, Nicole, had South African heritage, so this was a journey into the past for her. But we brought along the future, too, in the person of my son—and Nicole's soon-to-be stepson—Barney.

Johannesburg

Our loveLife road show trundled into "Jo'burg" in August 2002, wintertime in the Southern Hemisphere. We parked our yacht in the Ubuntu village near the Sandton Convention Center, amid a burgeoning assemblage of booths, exhibits, and signage. The *2041* was, safe to say, the only oceangoing sailing yacht at the conference.

The summit brought together 180 nations, 100 heads of state, 40,000 delegates. George W. Bush chose not to attend and was summarily treated as the summit's official whipping boy for his resistance to signing the Kyoto Protocol. The United States, paradoxically the world's most advanced country in environmental regulation and at the same time the world's most polluting nation, still was not a signatory.

Secretary of state Colin Powell appeared as the top American delegate. Hecklers interrupted his speech, shouting, "Shame on Bush!" The audience slow hand-clapped. The American stiffing of the greenhouse-gas accords was not popular.

Since the Rio Earth Summit ten years earlier, some of the faces had changed among the world leaders assembled, and the political landscape had thoroughly transformed. Tony Blair had succeeded John Major. Jacques Chirac replaced François Mitterrand. And Chancellor Gerhard Schröder of Germany was there, not Helmut Kohl. South Africa, the conference host, had held its first truly democratic elections.

The only faces held over from Rio were Robert Mugabe and Fidel Castro. Only dictators were the same.

I had the usual allotment of five minutes to speak to the world leaders and report on the global and local missions with which they had charged me at Rio. I am sure they had waited, chewing their nails anxiously in the intervening decade, worrying over what I would deliver—well, no, actually, I was quite certain they had totally forgotten their casual assignments that resulted in me busting hump for four years at Bellingshausen. I had made these promises at Rio, fighting tooth and nail and checkbook to keep them, and here I was to report on them. Only there was no one there to listen. I'm not sure Fidel and Mugabe remembered me.

Forgotten or not, I was determined to make a splash. I spoke about Bellingshausen, the global mission to clean up waste areas around the world. I moved on to a quick summary of our local mission with loveLife. Then the curtains behind me in the conference hall opened dramatically, and there, parked outside on the lawn, was the *2041*. It was a great moment.

Outside the conference hall I met with Jacques Chirac and South Africa's Thabo Mbeki. Mbeki smiled and nodded, shaking my hand. "Please carry on with your wonderful work," Chirac said in accented English. "We look forward to welcoming you to the next world summit."

I felt a little stunned at that. *Good Lord, French president dude, we've only just gotten to* this *summit, and already you're signing me up for the next one?* But all I said was, "Of course, sir, we'll be there."

And that was that.

Well done, carry on. I don't know what I expected. Maybe some bouquets, perhaps being lifted on the shoulders of world leaders and carried through the cheering audience of delegates. Yes, that would have been nice. I thought about the three million pounds of trash we had hauled out of Bellingshausen. How about three million pats on the back?

Don't expect applause. Another hard-won leadership lesson. Because if you do it for the applause, I realized, there are never enough hands in the world to clap for you. This idea had a flip side, though: Be sure to take the time with your team to celebrate its accomplishments. Life is short. Stop and smell the laurels.

I had been stuck on mile eighteen for so long, rushing from Footsteps to Icewalk to Rio to Bellingshausen to South Africa, that I had at times neglected to do this. As the summit wound down amid assorted volleys of political infighting, I made sure the loveLife crew and I embraced the fact we had just completed an amazing journey, reaching thousands of people for a good cause.

As I stood next to the *2041,* in the shadow of the summit conference hall, Doug Jackson, then the head of Coca-Cola South Africa, approached. He craned his neck to look up at the landlocked yacht. "You brought this boat all the way overland?" he asked. I told him that we had done just that, and sketched out the loveLife tour.

He said something then that changed my life. "Well, you can't just leave it here. You've got to sustain that inspiration, don't you?"

Sustainable inspiration. That's the first time I heard the concept codified in a two-word phrase around which I could wrap my mind. It was something I had been searching for, and Doug put it into words.

In the modern world, we are consumed by consuming. Everything is new. Everyone gets jaded. We consume not only products but enormous amounts of information, too. One message supersedes another with amazing rapidity. What matters gets lost in the chatter.

That's where the message of 2041 is different. Here was a concept, I realized, that was in it for the long term. It was an idea that a young person could dedicate herself to for a long time. Standing there at the yacht beside Doug, I calculated the years. At that point in time, the inspiration of 2041 would be sustained over the next thirty-eight years.

"You've done South Africa," Doug said. "How about we sort out the whole of Africa now?"

With Doug's support, we didn't just sail off into the sunset. We had taken the show on the road, and now we took it on the water, completing the Cape to Rio Yacht Race in January 2003. The group members had already spent a lot of time on the yacht—but only on land. Now they learned sailing and seafaring. The young sailors took the message from the second world summit to Rio, back to the site of the first world summit.

Their experience had created a team of young champions. The story of just one of the South Africans can serve as an example. Eric Bafo grew up in a poor, single-parent family in Cape Town's Khayelitsha township. His older brother was in jail. Eric himself appeared headed down the same track of crime, violence, and drugs.

His life changed almost by chance. On the dusty streets of the Cape Flats township, he followed what he told me were a group of "beautiful chicks" to a loveLife community center. "When I got there, what I saw was amazing. I forgot all about the women," he said. There was a basketball game going on, computers, people "talking my language."

"It touched me. I came out a different man. I wanted to be part of the solution and not part of the problem anymore." Eric took a sailing course offered out of the loveLife center in preparation to becoming a 2041 crew member.

This was the way to create young leaders. Under Captain Derek Shuttleworth, and again with Doug Jackson and Coca-Cola's support, in a yearlong voyage beginning in May 2003, we completed a full circumnavigation of the African continent. On-the-ground logistics were handled by Jason Whiting and my new sister-in-law, Laura Beyers. Nothing like getting the family involved.

Leaving from Mombasa in Kenya, we proceeded down the eastern coast, stopping at ports, performing cleanup projects on beaches, promoting clean-water conservation and recycling, doing our AIDS-awareness presentations for thousands of young people. As winter led into the heat of an African summer, the vision of Eric and the five other members of the loveLife crew in full Antarctic gear for AIDS-awareness presentations became ever more startling. At every step of the way, we were met with more crowds and further invitations to appear.

We stopped at thirty ports of call in all. In Maputo, Mozambique, Graça Machel, wife of South African statesman Nelson Mandela, wrote in the ship's logbook: "Congratulations for taking up the challenge to embrace the future, preserve and protect it. More importantly, for empowering young people to be the makers of history and the future."

From Maputo we sailed east to the island nation of Mauritius, the only known home of the extinct dodo. The stop in Port Louis, the capital, occasioned a symbolic pause for reflection: Could *Homo sapiens* possibly be going the same way as the dodo? The prospect appeared to be mere doomsaying, but then again, some of the climate-change models showed the average global temperature going up, up, up, to a degree where the planet would become inhospitable to humankind.

Doomed or not, the *Homo sapiens* aboard the 2041 continued around the Cape of Good Hope for a stop at Cape Town, of course, then north to Walvis Bay in Namibia and Luanda, Angola. Up the western coast, past the Straits of Gibraltar into the Mediterranean, across the coast of the Maghreb, through the Suez Canal, into the Red Sea, and back down the eastern coast to Mombasa.

Adrian "Jumper" Cross was an ex–Royal Navy seaman who helped tremendously with the circumnavigation effort. Doug Jackson saw to it that the Coca-Cola company in Africa came on board, literally, as a sponsor of the yacht and the voyage. The effort was managed by the indefatigable Nigel Holman.

Coke sent employees as crew members on sea legs from the various ports of call along the way. AIDS prevention was not an abstract cause for Coca-Cola Africa. It was a life-and-death issue. As Africa's largest employer, the company was seeing its people die from the disease in distressing numbers.

Getting to know the inner workings of such a massive business as Coca-Cola came to be one of the true privileges of 2041. It heartened me, since I realized that corporate culture was slowly changing, reorienting itself to new realities. I learned of this primarily from one of Coke's environmental point people, my American "brother," Jeffrey Foote.

Under the initial leadership of Doug Daft and Brian Dyson, Coca-Cola was among the first businesses to recognize that conservation of water and packaging, plus the use of alternative energy, meant that green = green; that is, conservation and alternative energy meant cash savings on the bottom line. It also showed leadership in corporate social responsibility. Coca-Cola's work in this area has been built upon by current executives Neville Isdell and Muhtar Kent.

The Coca-Cola and bottling company employees we worked with on the circumnavigation put human faces to the worldwide soft-drink brand. "Coca-cola" is the second-most recognizable phrase in the world, behind "okay." In turn, the company helped raise our visibility on the *2041*.

An epic journey, all told. The crew members from loveLife became the first Africans on record to complete a circumnavigation of their home continent. More than that, they matured into champions of causes in their own right, going on to participate in AIDS initiatives (such as Zackie Achmat's Treatment Action Campaign) and promoting environmental causes (Abahlali baseMjondolo's clean-water actions).

I recall a special moment when we returned from the Cape-to-Rio race. The *2041*'s ragtag crew of eight young students were invited to lunch at the Royal Cape Yacht Club in Cape Town. It was the kind of place that they, as black men, would not have been allowed inside (except as servants) just a few years earlier. As we sat there, a puffed-up yacht-club regular, complete with a blue blazer and a captain's hat, leaned over to our table and addressed Sithembele "Joe" Cata, one of our crew members. "Well, young man, and how many nautical miles do you have under your belt?"

That's what old sailors do, in idle hours around yacht clubs, ask one another how many nautical miles they have and compare the numbers for boasting privileges.

"Thirty-four thousand," Joe answered.

Mr. Blue Blazer blinked. That number represented a lifetime of boating. "And how long did it take you to rack up thirty-four thousand nautical miles?" the patrician regular asked.

"Two years," Joe said.

Blue Blazer sagged back in his chair. Conversation over, advantage Joe. Brilliant. Ten years later Joe joined the Royal Cape Yacht Club as a full-fledged member. Through the efforts of young people like Joe, we got the message out to all of Africa about the preservation of Antarctica.

After the Africa circumnavigation and the Cape-to-Rio race, the *2041* continued to prowl the world. I felt I had something to prove. I wanted to show that alternative technologies could stand up to the harshest conditions on earth. We went looking for exactly those kinds of conditions, and in 2004 in Sydney Harbor, we found them.

Sydney to Hobart

Plastic refuses to decay. A plastic bottle dumped into a landfill will just sit there, unrottable, immune, pretty much forever—or at least for one thousand years.

That's the time it takes for an average, ordinary plastic water bottle to degrade. And we throw away about forty million of them every day—out of some thirty billion sold in the United States every year.

The accepted wisdom on reusing plastic is that recycled stuff is just not as strong and durable as the original. An urban myth tells the tale of lawn furniture molded from recycled plastic melting in the sun.

I wanted to do what I could to refute the myth. "What if we could make our sails on the *2041* out of recycled materials?"

It was the kind of random, wondering-out-loud what-if that the people around me have come to dread. *What if we could walk to the South Pole? What if we could haul a yacht overland?*

Once again, John F. Kennedy is relevant here. He committed the United States to go to the moon. He announced a ten-year deadline. Of course, Kennedy himself didn't have to go near a space capsule or a moon rover. He didn't have to figure out how anyone else could do it. All he did was toss out a what-if and inspire other people to accomplish what he had promised.

We would fashion our sails out of recycled plastic bottles, and we would use them to fly the *2041* to the moon. Well, no. But we would use them in the next-closest thing, the famously brutal Sydney Hobart Yacht Race, an annual dash across some of the roughest seas on the planet, a race that was nicknamed "the Everest of Yachting."

Doug Jackson, then president of Coca-Cola's Southern and East Africa Division, put us in touch with his Australian counterpart, Mike Clarke, who loved the idea of making the sails out of Coca-Cola PET bottles.

The problem was, I hadn't the foggiest notion how to make sailcloth out of plastic bottles. I didn't know anything about elasticity, tensile strength, UV resistance, flex loss, or creep. I was ignorant of weaves, films, and adhesives, and about Kevlar, PEN, and carbon fiber sails.

What I did know was people, so I contacted one of my Immortals, Christine Gee, now partnered with my sobriety savior, Australia's best-selling author, Bryce Courtenay.

Working from her home country of Australia, Christine had been hugely instrumental in putting the Icewalk expedition together. I knew from experience that I could give Christine a task and she would follow through to the finish with elegance and style. She was an absolute terrier. But at first I thought that in proposing to use sails of recycled materials, I had bitten off more than even Christine could chew.

While polyethylene terephthalate, or PET, was a common material utilized for modern sails, no one had ever really made sailcloth from the recycled PET that came from bottles. A ripped or shredded sail can be lethal. With images of melting lawn furniture dancing in their heads, sail manufacturers did not want to take the risk.

Christine would not take no for an answer. She connected with an Australian sail company called Dimension-Polyant. The idea was to use the recycled PET sandwiched into a D4 membrane between taffeta outer surfaces. We needed a strong, durable material that wouldn't shred in high winds.

The kind of short-fiber PET produced by most recyclers was totally unsuitable. Christine had to journey to Japan to find companies that made the right stuff: Teijin Group and Marubeni Corporation. Dimension-Polyant and a company called Ullman Sails took it from there, with financial help from one of the world's largest recycling companies, Visy. The 2041 had her brand-new recycled sails in time for the start of the Sydney Hobart in December 2004.

In 1987, in the course of organizing for Icewalk, Christine and I had stood on the shoreline of Sydney Harbor, watching the start of that year's race. It was an incredibly romantic sight. Under the crystal-blues skies, the various yachts jockeyed for position at the starting line for the 626-nautical-mile competition. I turned to Christine. "Someday, we'll have a go at this."

Seventeen years later, we did. No one knew if the recycled sails would work. But Sydney Hobart definitely represented an extreme test case.

The race, traditionally begun on Boxing Day, December 26, sent a regatta of world-class racing yachts out of Sydney Harbor, down the eastern shore of Australia, and through the Bass Strait to Hobart. These are harsh, changeable seas, where the prevailing westerlies and polar easterlies mix and collide with the Australian landmass. For added relish, the waters are the notorious haunt of the great white shark.

In the tragic 1998 Sydney Hobart race, fierce storms battered the course. Out of 115 yachts that started, 44 made it to Hobart, 5 boats sank, 66 boats retired, and 6 sailors died. Uncannily, that race started with brilliant sun shining in Sydney Harbor.

Brilliant sunshine also marked the beginning of the 2004 race. Coca-Cola, with its active interest in the possibilities of recycling plastic bottles, supplied not only sponsorship but also crew members for the *2041* under the captaincy of Bret Perry.

We had another Brett, a schoolteacher, Brett Austine, who had won an award for innovative work with young people, with the prize of a berth on the boat. The rest of us were more or less experienced sailors. Fraser Johnson was our ringer; he had been involved in thirty previous Sydney Hobart races, even though the event had only been running for fifty-nine years.

The yacht was temporarily renamed, for the sake of the race, the *active factor,* after an all-lowercase Coke slogan emblazoned on its recycled sails.

There is no more thrilling start in yachting, and maybe only a few in any sport, than the beginning of the Sydney Hobart race. Massive ninety-eight-foot super-maxi-class yachts, big cruisers, elegant cutters, and ocean racers surge across the starting line in ranks, within view of the soaring opera house on the shore.

The *2041* was among them. I didn't much like the weather forecast, especially for day two: gale-force winds, heavy seas, extreme cold, even hail.

The occasion represented the sixtieth anniversary of the race, so it had attracted quite a few boats. But out of a field of 116 boats, 40 dropped out ("retired at port," in the polite language of yachting).

We pondered quitting ourselves. The Clash song ran through my head: "Should I stay or should I go?" Sunny Sydney Harbor was safe and within striking distance of Barney and Nicole on shore. I pictured our newfangled, untested sails shredding, the yacht adrift under bare poles, great whites circling, helicopters hovering to rescue us.

Luckily, I had a great crew. The elder J. P. Morgan once said, "You can do business with anyone, but you can only sail a boat with a gentleman." Those aboard the *2041* on the Sydney Hobart were a band of brothers and sisters, everyone committed to the race and excited to be there. The engineer and foredeck hand, Inigo "Ini" Wijnen, a thirty-four-year-old Dutch jack-of-all-trades, had overseen the refitting in Australia and could hold the boat together if anyone could. Ini believed in the sails. Bret Perry, our sure-hand-at-the-helm captain, believed in the sails.

Should we stay or should we go? We went.

We were twenty tons heavier than most of our competitors. After all, our boat was an ice yacht, specially strengthened for polar waters and pack ice. They were greyhounds. We were a Saint Bernard.

It seemed difficult to believe, that sunny first day in calm waters, that any cloud would mar our horizon. But then extremely troubling news came over the radio.

"Rob," Ini called to me. "Listen to this. Something big is going on."

A tsunami was reported in Asia. All that day, more news came in and the mood darkened as the true extent of the tragedy became apparent. The first bulletins said, "A hundred persons injured." Then, "a hundred deaths." Then a thousand deaths. Then ten thousand. A hundred thousand. A massive catastrophe.

Meanwhile, it was a bright sunny day on the ocean. The disconnect jarred us all. But in front of us, directly in our path, a black sky loomed ominously.

On day two, it hit. *Wham!* We piled into a massive storm. We "shipped it green" as the angry sea swept our decks. It was the worst storm I have ever experienced, on sea or land. Fifty-knot winds, a full gale, tumbling, twenty-foot waves white with foam and spindrift, each one slamming into us with heavy, shuddering impact. We were doing well at the back of the pack. The *2041* was designed for that kind of weather. I became seasick for the first time in my life. I had always looked on with secret contempt as landlubbers lost their lunches in heavy seas, but now that I was suffering, I saw things differently. I wondered if there was a pistol on board, so I could put myself out of my misery. Brett, the schoolteacher, and I lay in our racks, sick as dogs, he in the top bunk, I groaning and retching on the bottom. It wasn't the easiest moment in my life. I was thinking I should have been back with my son and wife.

We kept hearing bulletins as one yacht after another dropped out of the competition. Aussies don't give up easily. They would need an overwhelming reason to leave something as high-profile and long-prepared-for as the Sydney Hobart. The famous race suddenly felt more tenuous, a mere collection of wood chips on a lethal ocean.

But it was exactly here that the *2041* showed her mettle. She was built for these kinds of seas. We started moving up in the ranks, passing one boat after another.

Then another bulletin, on channel 16, the emergency frequency: The *Skandia,* the super-maxi leader of the race, had snapped a keel and capsized. Captain and crew took to life rafts and were picked up by rescue helicopter.

We forged on. The sails held. The *2041* proved herself to belong in one of the most competitive yacht races in the world. We crossed the Derwent River finish line twenty-fourth out of a remaining field of fifty-six, and cruised triumphantly into Hobart Harbor. It was like a Land Rover beating out a fleet of Formula Ones.

Case proven. Old Coke bottles rescued from the landfill can make damn good sails. Recycling works, even in the Tasman Sea. Every other sail involved in the race would be dumped into landfills. Ours could be recycled again and again.

Over the next few years, I again would attempt to test the worth of alternative technology in even harsher conditions, by doing something no one else had been able to do.

We would homestead in Antarctica.

E-Base

During the Africa circumnavigation, and in the middle of the Sydney Hobart race, I sometimes felt as though I was cheating on Antarctica. I loved Europe, I loved the Americas, Asia, Australia, and Africa. But I was wedded to another continent.

Always, whether the *2041* was docked in Boma in Zaire or plunging through big seas in the Bass Strait, we were getting out the message of *2041*. A peculiar vision floated in my mind's eye. A small, rectangular outbuilding, trailerlike, painted rust red, perched on the slopes above Maxwell Bay at Bellingshausen Station.

I first noticed the place during the Bellingshausen cleanup effort. It was empty, the Russians told me, nothing more than a hut formerly used for communications but now abandoned. It proved bleak and drafty during my initial visit. If the hut were left alone, the steady winds that blew out of the south would soon flatten the whole building.

The site perched on a bony spine of the Fildes Peninsula above Bellingshausen, situated at 62°10' S latitude, which is roughly analogous to the Northern Hemisphere's Scandinavia. With a view looking south across Collins Harbor and Maxwell Bay toward the pole, the hut was removed from the Russian station itself. Lonely, isolated, magnificent.

As soon as I entered, a sense of déjà vu flooded me. It was as though I knew the place well. The interior was about the same square footage as our In the Footsteps of Scott hut at Jack Hayward Base. Outside the windows I saw a view similar to that I had seen back then: volcanic beaches, dark, ragged hills, and the slate-gray waters of a polar bay. I had come full circle.

Private citizens are not allowed to "own" a building in the Antarctic. It's just not done. *Don't even think about it, Robert.* But it turns out that I could not *not* think about it. I could not get the little red building out of my mind. The hut at Maxwell Bay, I decided, would suit the purposes of the 2041 effort very well. Besides, what did "ownership" mean? No one owns anything in Antarctica. Maybe my ownership of the hut would be more in the nature of custodianship.

I continually fretted over whether I was being part of the solution or part of the problem. In 2003 we began taking groups of students and business leaders on trips across the Drake Passage from South America to the Antarctic Peninsula, calling the initiative Inspire Antarctic Expeditions (IAE).

Antarctic tourism was exploding. That year tourist visitors to the continent totaled seventeen thousand. (Five years later, in 2008, the number had doubled.) Was this a good thing or a bad thing? Wasn't it a violation of the sage vision of my mentor Peter Scott? "The world ought to have the sense," Scott said, "to leave just one place on earth alone—Antarctica."

Those were his words. Would Sir Peter have approved or disapproved of the influx of Antarctica visitors? Sadly, he was not around for me to ask. If he meant that no one should go to Antarctica, then I was afraid I could not agree with him. But if, as I thought was the case, he meant there should not be exploitation of Antarctica's resources, I could not have agreed more.

Dozens of ships cruising the waters of the peninsula, the Weddell Sea, McMurdo Sound, and the Ross Sea, disgorging hundreds of passengers for forays into the fragile Antarctic ecosystem—could that possibly be good? Of all land species, only a herd of elephants exceeds the damage done to a landscape by a troop of humans.

But I recalled a signal lesson taught to me by Steve Irwin, "the Crocodile Hunter." Wildlife, he said—and I suppose his point could be extended to wild places—are in direct competition with thousands upon millions of other attractions, images that blast at us in today's rapid-fire culture. The Internet, TV, video games, downloadable music, all the multiplying data and input battle for eyeballs. The telephoto-lens approach to wildlife just won't suit anymore. As virtuous as it is, from a Peter Scott "leave it alone" point of view, the distancing factor decreases the impact. During his life, Irwin was oftentimes assailed for getting viewers up close to his beloved crocs, venomous snakes, and monitor lizards. If we are going to save these animals, he responded, we have got to allow people to get to know them.

Jacques Cousteau put it another way, giving the rationale behind his TV documentaries aboard his famous craft the *Calypso:* "People protect and respect what they like, and to make them like the sea, they should be filled with wonder as much as informing them."

I noticed that the Antarctic tourists I came into contact with were just as enthusiastic about the place as the scientists. Add in the fact that the tourists don't scar the landscape with permanent bases and trash dumps, and you can make as fair an argument for tourism as you can for science.

The Antarctic expeditions I ran represented a clear compromise. No, they did not "leave Antarctica alone," as Sir Peter Scott would have it. But mitigating their effect on the landscape was the much greater impact (I hoped) on the trend of public opinion.

We carbon-credited every expedition, so as to leave as small an imprint as possible. I tried not to cater to the common tourist impulse of itinerary travel, which involved going to a place merely to tick it off of a been-there-done-that list. I wanted to create ambassadors—especially future ambassadors—for 2041. I wanted to fill visitors with the sense of wonder that Cousteau cited.

With the support of Coca-Cola Europe's Sandy Allan, we hit upon a combination of business leaders and students for my annual IAEs. Perhaps nowhere else in the world could an engineering student from Temple University in Philadelphia get to pick the brain of a vice president from British Petroleum, or the VP catch the sense of enthusiasm and excitement from an undergrad. It was cross-pollination at its best.

Over the course of rigorous, two-week expeditions in Antarctica, we hosted climate-change workshops, sustainable-leadership workshops, presentations by alternative-energy gurus.

The IAEs gathered my Immortals together and added in a few more. Peter Malcolm was at my side once again, as was Adrian "Jumper" Cross, who handled safety and training. But the real saving grace in my life, and the CEO, prime mover, and guardian angel of the 2041 organization, was Anne Kershaw, the widow of pioneering polar pilot Giles Kershaw. Annie was a godsend in organizing the complex logistics of 2041's annual IAEs.

But, instead of bringing people to Antarctica, what if there was a way to bring Antarctica to the people? Three things were essential for my 2041 effort: education, education, and education. What if I had my own base in Antarctica, from which I could reach out to classrooms, communities, activist groups, and businesses around the globe?

That's where the small red building overlooking Maxwell Bay at Bellingshausen came in. I petitioned the Russian Antarctic Expedition (RAE) to use the hut for the creation of the continent's first base dedicated entirely to education. Not a scientific-research station—there were already more than a dozen of those—but one devoted to educating young people about life at the bottom of the world. Since I had built up a lot of goodwill via the Bellingshausen cleanup, the RAE graciously acceded to my request. Coca-Cola, under the leadership of Andy Duff, and RWE N-Power from the United Kingdom came on as sponsors.

E-Base, short for "education base." That's what I quite grandly named the little hut, which given its disrepair was a bit like calling an outhouse "O-Base." The place needed a lot of work before it would approach the realization of my dream.

My idea was simple. Rebuild the hut into an accommodation unit, and using recycled materials build a new education base that ran entirely on renewable energy. That meant solar, wind, and geothermal. If we could make it happen here, in the harshest environment on earth, we could use our efforts as a template for other, more hospitable parts of the planet.

We could trace a history of energy use in successive Antarctic expeditions. Captain James Clark Ross lit his lamps on *Terror* and *Erebus* with whale oil and used wind power to force his way through the pack ice to reach the barrier for the first time. Scott's original 1910–12 *Terra Nova* expedition used primarily Welsh coal. In 1985–86, In the Footsteps of Scott used primarily diesel fuel and Welsh coal. Now, in the new millennium, E-Base would use wind and solar thermal energy.

Whale oil to coal to petroleum to renewable energy. A blueprint for the future. If we are using renewable energy in the world, we won't have to exploit Antarctica in the future.

Once again, I didn't quite know what I was getting myself into. The holy trinity of sustainable living—reduce, reuse, recycle—applied only unevenly to the effort to get E-Base up and running.

Utilizing as many resources as possible from within a community is a cornerstone of green design. But Antarctica had no manufacturing base, of course, no building materials produced or indigenous supplies. Many of the products we needed had to be imported, adding fuel costs and hydrocarbon emissions to the project. We did the best we could. All of the building materials we chose for E-Base were sustainable products. We brought in structurally insulated floor panels, 100 percent postconsumer recycled rubber interior flooring, and an Energy Star watertight blanket for the roof and siding.

In March 2006, in some of the worst conditions King George Island could throw at us, with alternating freezing sleet and high, blustery winds, we started work. We were greeted at the site by Alejo Contreras, 2041's man on the ground in Antarctica, last seen at the fuel pumps in the Patriot Hills.

In the years since then, Alejo had probably logged more hours on the continent than any other living person, mainly at Chile's Frei Montalva Station near Bellingshausen. He had been instrumental in getting Gareth and the boys out from Jack Hayward Base in 1987 and had been threading in and out of my life since then, a true Immortal.

If I am ever down or depressed in Antarctica, fretting over this difficulty or that screwup, it often helps for me to talk to Alejo. If he's not present, merely thinking about him lifts my mood. I see us from his eyes, and I have to smile. For some polar wannabes, we must resemble a group of fumbling, unlikely Don Quixotes, tilting at windmills. But Alejo understands.

In the case of getting the E-Base established, the metaphor turned into the literal truth, and tilting at windmills almost wrecked the whole endeavor.

Light

―――――

E-Base acted as a tool, the same as our sailboat, the *2041*. I liked to think of them in purely symmetrical terms. Where were the two areas of energy consumption most people encountered on a daily basis? Home and transportation.

On the *2041* (the transportation part of the equation), we reengineered our systems for alternative power, until the yacht was a totally green entity. In setting up E-Base, we sought to accomplish the same style of reengineering on the home front, transforming domestic energy use for heating, hot water, and light in the least friendly environment on the planet.

In a sense, 2041's annual IAEs, the expeditions we took to Bellingshausen and the peninsula every year, were planned with the E-Base in mind. One of the purposes of the IAE trips was to get the E-Base up and running.

Each year, on successive IAE trips, we pushed the effort forward. On expeditions one, two, three, and four, we dug holes and laid out new foundations. On the fifth IAE, in 2007, we did all our new construction. My team on the E-Base project included the great polar photographer John Luck and 2041's superb videographer, Kyle O'Donoghue, so at least our battles were well documented.

On the sixth IAE, in March 2008, we tested out the base with our new alternative-energy systems. Crammed into the E-Base upon arrival, we first had to erect our living quarters, an "E-Home" tent snugged up against the northeast side of the hut. Made by Winterhaven, the E-Home was shaped like a World War II–era Quonset hut but constructed out of space-age materials. It kept us out of the wind, but with no electricity it was, as John Luck put it, "a dark cave."

To be civilized is to require light. Put another way, civilization *is* light. If you don't subscribe to this particular belief, attempt to do without it for a few nights. Our immediate task, attempted in the face of a wet, pelting snow, was to rig our wind generators. They would give us electricity, if not enough to supply all our heating needs, then enough to light our weary nights.

Wind we had, in abundance. It soon proved to be the case that we were in fact overblessed with it. Unrelenting Antarctic winds howled day and night across the ridgeline. The strong gusts blew apart one of our turbines, the one we had named Nemo after Verne's intrepid hero—or perhaps after the cartoon clown fish.

Our other generator, Charlie, consistently locked in the "off" position, a built-in feature that was designed to shut it down in gale-force conditions but not helpful on the Fildes Peninsula. Charlie's tail fin furled so completely that the turbine parked its nose cone permanently facing away from the wind.

It was grim. Although Antarctica is traditionally pictured as a winter wonderland, many times it is much more of a muddy mess. Only 2 percent of the continent is not under ice, but at times it seemed that every percentage point of that exposed dirt was at E-Base. Freezing mud got tracked inside, infiltrating into every nook and cranny, mucking it all up and making cleanliness next to impossible.

The weather remained a constant malignant force, wearing down the team. The winds gusted to twenty or thirty knots. We were cold, wet, and disheartened by our difficulties with the generator. I thought of Voltaire's purported last words: "More light!" I hoped they wouldn't prove to be mine.

Reduce, reuse, recycle. We could not trot off to the nearest hardware store to purchase a part in order to rejigger the turbine to prevent it from shutting off. We scavenged the hills around the dark and powerless E-Base, trying to turn up a scrap of metal, a bit of wood that would help us. We needed a simple length of steel rod to attach to the tail fin of the generator, to block it from going into a parked position.

Johnny Luck, the irascible photographer, was fond of quoting what he said was an adage of the Great Depression: "Use it up, wear it out, make it do, or do without." Mark Nicol and Russell Oliver from Britain's N-Power worked in constant wind and blowing snow to refit the generator, "making it do."

Others on the team, among them Nicky Wootton, our link to sponsors and magician on the Apple computer, and Marjan Shirzad, in charge of communications, scavenged supplies the best they could. Plastic loading pallets left behind by a resupply ship to one of the international research stations in the area formed the floor of the E-Home. A rusting pile of decades-old angle irons and steel cable became, with the help of a hacksaw and some physical effort, a full set of tent stakes and guy lines that secured the E-Home against strong storm winds.

Dwight Eisenhower used to say that in his experience of battle, plans were useless but planning was indispensable. We had planned to be up and running at the E-Base site almost immediately, snug in our Winterhaven tent, with two Mongolian yurts erected on the other side of E-Base for supply warehouses and additional quarters. One day stretched into three.

No light, no heat, no power from the generators.

Let us now praise fossil fuels. Petroleum, coal, natural gas. They have given us so much. They have made an era of cheap energy possible, and we have exploited them mercilessly to provide all the comforts and blandishments of civilization.

It demands a peculiar kind of bad faith to sit in a heated house in front of a computer made largely of plastic, writing angry letters to the newspaper about how we are all going to hell in a handbasket because of that old devil oil. Afterward climbing into a car to head off to a protest demonstration against Exxon, driving over asphalt roads or mighty concrete interstates where each intersection is thoughtfully well lit by mercury-vapor lamps.

During In the Footsteps of Scott, there were times when my eyes teared up and my heart filled with something very much like love at the sight of the stove's yellow-blue flame. When you are in a tent in Antarctica and it's −40 degrees F outside, the hiss of the stove burning its fossil fuel is the most welcome sound of all. The wavering heat it puts out means everything to you—life, liberty, the pursuit of happiness.

All our kit was to a greater or lesser degree the product of hydrocarbons. Sometimes this was literally true, as in the case of the polymer fabric of the tent itself, manufactured by heating a crystalline salt of carbon, hydrogen, nitrogen, and oxygen to 545 degrees F. Other times the dependency on fossil fuels was more removed, for example, the sausage we ate in the tent, which had been on the hoof at some point in its distant past, but was nurtured, manufactured, and transported via petroleum-based processes.

Scott died out on the Great Ice Barrier not because he had no food, but because his fuel supply ran out. The gaskets beneath the caps of his fuel containers did not seal correctly in the intense cold, so the paraffin for his stoves leaked out and evaporated—one of Antarctica's first environmental disasters. What fuel remained proved insufficient to cook his food and melt snow for water. The bottom dropped out for Scott when he arrived at the Middle Barrier depot on his journey home from the pole and found the fuel cache there "woefully short."

Our own global predicament today is Scott's Middle Barrier depot writ large. I often try to imagine the awful sinking feeling he must have had, lifting a leaky fuel can and calculating by its weight that it was half-empty. He might have smiled gamely to Bowers and Wilson, but fingers of dread and panic must have gripped his consciousness.

I catch a whiff of the same sensation standing in Times Square or Piccadilly Circus, amid the internal-combustion rush of cars and trucks, knowing that the end of the Age of Oil is near.

Stored sunlight. That's all that fossil fuels really are. His Grace, as Amundsen called the sun, irradiates the planet, creating a biotic mass that certain natural processes render into coal, petroleum, or methane-heavy natural gas. When we drive to the store for a quart of milk, we are burning sunlight that fell upon the earth millions of years ago. Stored sunlight consumed, in the case of oil, at a rate of thirty billion barrels a year.

How long can we suck on that teat? How long before we find ourselves in Scott's predicament, in a cold tent shaking a half-empty fuel container, a cosmic "uh-oh" forming in our collective consciousness?

So, yes, fossil fuels are fantastic, fossil fuels are wonderful, fossil fuels are the greatest thing since sliced bread. But their day is about to be over, and it's time for us to move on—the sooner, the better. Not the least because burning them is about to render our planet uninhabitable.

On both Icewalk and In the Footsteps of Scott we used a crude form of renewable energy as a backup for our gas stoves. We traveled with black plastic bags (made of petroleum-based polymers, but that's a quibble) that we filled with snow. The weak polar sun would irradiate the bags and melt the snow. At E-Base, we created and made do and improvised as we went along. That coping quality is oftentimes a characteristic of renewable energy. When a wind vane jams shut, you have to be able to come up with an alternative on the spot. The fix might not look great, but it has to work or you'll remain cold.

A black plastic sack full of snow is just not as elegant as an MSR XGK camp stove, with machined brass fittings and a blued-aluminum fuel reservoir. One is simply cooler than the other. It can't be helped. There is something ungainly about solar panels, too, especially when compared with a buried pipeline delivering natural gas.

Some alternatives to fossil fuels are going to be messier, less sleek, more improvisational, and downright shambolic in comparison to what we are used to. It's just another hump we have to get over on our way to energy independence.

The Valley of the Shadow

In full view below us at E-Base were two international research stations, Russian and Chilean, run by fossil fuels. At any time, we could have slipped the leash and gone down to Bellingshausen to get warm, maybe even enjoy the treat of a hot shower. The Russians would have welcomed us. Who would know? No one but ourselves. The pull to head "downtown" was at times overwhelming. It was a measure of our impoverished straits at E-Base that even the hardscrabble conditions at the Russian research station appeared luxurious.

The lure of Bellingshausen represented, for me, the constant temptation to compromise on the principle of going green. Why endure all the hassle with wind generators, solar panels, reducing and reusing and recycling, when it was so much easier to plug in to the grid or jump into the car?

But there was a flip side, literally, to the E-Base experience. Down the slope to the east, in the opposite direction from Bellingshausen, stretched an empty expanse of pure wilderness. I used to think of it as the Valley of the Shadow—barren, unwelcoming, almost lunar in its rocky promontories. If the Russian base was a compromise, then the valley represented a challenge.

I used to stare into it and think of Psalm 23. But also of Tennyson writing about British cavalry running into Russian gunners, perhaps the ancestors of some of the people down at the base.:

> *Theirs not to reason why*
> *Theirs but to do and die*
> *Into the valley of Death*
> *Rode the six hundred.*

I had seen this face of Antarctica before, of course, on Beardmore Glacier and elsewhere. The beauty and terror of the continent were often so close together as to be synonymous. The most beautiful person on earth, right at the moment when the mask was pulled off . . . Beauty, wrote Rilke, is merely "the beginning of terror" and "serenely disdains to annihilate us."

It was this aspect of Antarctica that I sought to honor by persevering in my efforts. I wanted to gain a foothold on the continent, one that would be powered wholly by renewable energy. When I considered E-Base from the Valley of the Shadow, it appeared insignificant, puny, laughable. The valley didn't care. It would be there long after I was gone.

But insufficient measures like E-Base were the only thing I knew how to do. I couldn't wave a magic wand that would help forever preserve Antarctica, the land of thousands upon thousands of such stark raving beautiful valleys and desolate landscapes. I could only put my shoulder to the wheel in this particular fashion.

Once again, it was Em who brought it all home to me. "The only way we will save Antarctica is if there's no reason for anyone to go down there and destroy it," she said, in response to my second thoughts and self-doubts. "That's your reason for your E-Base, darling, and all your alternative-energy work. Make it so there's no reason to go down there and drill and mine."

A few hundred yards from E-Base, perched along the same ridge-line, the Russians had erected the magnificent Trinity Church, a tiny Russian Orthodox gem constructed of Siberian pine, built in the mother country, disassembled, shipped south, and reconstructed there. The church, with religious icons of Jesus and Mary gracing its tiny interior, represented one answer to the looming Valley of the Shadow of Death, perhaps a much more powerful response than E-Base could ever hope to be. But for me, the most salient feature of the interior was not the icons but the heavy-weight iron chain that held the walls in place against the heavy winds.

It reminded me of a saying that Ronald Reagan famously quoted to Mikhail Gorbachev during their disarmament talks: "Trust everybody, but cut the cards." At Trinity Church above the valley, the motto was rendered a little differently. Trust God, but chain down your church.

I remember the moment when the light came on in the E-Home tent. It resembled a benison. Light, and not the kind of light that came from fossil fuels, not the kind of light that was slowly strangling the land that I loved. Light without a carbon imprint, without guilt, without tacit complicity in climate change. Because we equipped the E-Home with energy-efficient fixtures, the whole tent was lit, and well lit, by the wattage required by a single old-fashioned incandescent bulb.

Slowly, over the next week, we cobbled together the elements of E-Base's renewable-energy plant. Because sunlight in Antarctica is often weak and filtered through cloudy skies, conventional solar energy from silicon-based cells seemed ill suited.

Under the leadership of its founder and then president, Ed Stevenson, and founder/chairman Bob Hertzberg, a U.K. company called G24 Innovations provided us with panels made up of dye-sensitized thin-film solar cells, extremely flexible and capable of producing energy in low-light conditions. I saw the future of solar in G24 at E-Base, and it worked.

G24 Innovations had its main manufacturing plant in Cardiff, Wales, a massive 187,000-square-foot facility in the port where the *Terra Nova* and *Southern Quest* had both stopped to pick up coal. I felt the march of progress in Cardiff's transformation from a fossil-fuel center to a solar-energy hot spot. Amazingly, we had visited Cardiff for coal on the Footsteps expedition, and here we were in the next century visiting the same town for renewable energy.

E-Base came together. True to our sustainability credo, we framed the building's solar water-heating system using scavenged scrap wood. Russell Oliver reused a discarded futon bed frame, transforming it into a wall-mounted desk. We affixed a satellite dish atop the hut (carefully securing it with guy wires in the face of the constant wind), placed my camera-equipped laptop on the desk Russell had created, and on March 10, 2008, E-Base went live.

That we could connect to the world from Antarctica was largely a tribute to the work, energy, and enterprise of Marjan Shirzad, Iran-born, Brooklyn-raised, who proved herself a communications wizard. Anyone who has struggled to wire a home computer network can well

appreciate the insane complexities Marjan had to overcome in the effort to hook us up at E-Base, where bottom-of-the-world satellite coverage can be spotty and the power was supplied by wind and sun.

As always, we struggled with funding, too. I recall the British-born
Nicky, in the middle of the rolling chaos of E-Base, tugging on
my sleeve to get me to call that sponsor or that company. It's the constant lot of the expedition organizer. Nothing ever comes simple in
Antarctica.

Getting connected to the BGAN—the Broadband Global Area Network—was, for me, as emotional as the "let there be light" moment
when we first had wind-powered electricity. Here was the purpose, here
was the reason for all the mud, sweat, and tears. E-Base stood for
"education base," sure, but it also meant "electronic base."

Sitting in the still-chilly hut, I video-chatted with assemblies of
schoolchildren from Britain, some nine thousand miles away.

"Hi, my name in Kendra Kelsey and I would like to ask the question whether it is difficult to breathe where you are in Antarctica."

On my laptop, I could see Kendra. She could see me on her school's
video monitor. "It's not hard to breathe where we are now," I said.
"But it can be very hard and very dangerous just to breathe sometimes
in Antarctica."

I talked about how, in extreme temperatures of −80 degrees F or
below, exterior air can freeze teeth, cracking the enamel. "We wear fur-
trimmed hoods in front of our faces, so when we breathe out, we actually warm up the air we are going to breathe so it doesn't hurt us."

After E-Base went live, we had dozens of these intercontinental
video chats, connecting with schoolchildren, community groups, and
interested parties as far afield as India and China. I spoke on my trusty
Nokia phone with batteries recharged only by renewable energy. I
trusted these chats would help the cause.

Would a ten-year-old girl such as Kendra Kelsey grow up and
somehow influence whether or not Antarctica would be protected in
2041? There was no way to know for sure. There was a chance she
might, but if we didn't try, it was likely she would not.

I want to give one example from my own experience from a young
life transformed. During Icewalk, one of those twenty-year-olds in the

group who had gathered at Eureka was Peter Hobart, who had been living a carefree life as a bartender. He wanted to be Tom Cruise in the movie *Cocktail,* although his girlfriends thought he was better-looking. His father was worried about him, so he persuaded Peter to join the Icewalk group.

On Ellesmere Island, Peter experienced the splendor and desolation of the Arctic. He went to the workshops. He listened to Dr. Ian Sterling of Parks Canada talk about toxins in the fat of polar bears. He helped clean up fifty-gallon drums from waste dumps.

I had no clue whether the student component of Icewalk would have any effect at all. My basic premise resembled that of the American lottery advertisement: You've got to be in it to win it. I thought perhaps the Eureka experience on Ellesmere Island might inspire a few of the twenty-two students from fifteen countries to carry on the work of preserving the polar regions from exploitation.

The bet paid off, big-time. The students who gathered on Ellesmere Island in 1989, during Icewalk, are in their forties today. They have become business leaders and influential figures. Peter Hobart has continued to be involved in polar advocacy and 2041. He cites the Ellesmere Island experience as crucial in shaping his life. He is an assistant district attorney in Philadelphia today and still a huge part of my world.

Sitting in front of my laptop at E-Base, talking to Kendra Kelsey, I believed it truly was a wonderful world. I was linked up with a schoolchild a world away—another potential Peter Hobart. Was that not a fabulous development? It helped me have faith that if I kept at the effort long enough, I could reach enough people, create a few champions for Antarctica, and make 2041 a year of decision.

I needed that faith. Otherwise why had I frozen my butt off at E-Base, why was I perched on the edge of the Valley of the Shadow, why had I trekked and starved and presented and lectured and advocated? What was any of it for, if not to help Kelsey Kendra and others like her make the Antarctica connection?

E-Base was one way to get the message out. After spending three weeks living on renewable energy in the harshest environment on earth, we immediately embarked on a campaign to take the message directly to the decision makers, via a round-the-world voyage in the *2041.*

Voyage for Cleaner Energy

After my experiences at E-Base, I became more convinced than ever that, paradoxically, the future of Antarctica lay in the temperate zones. The smokestacks and exhaust pipes of Los Angeles, Mumbai, and Shanghai were the biggest threat to the effort to preserve Antarctica as pristine wilderness. In order to save Antarctica from being exploited for its resources, the only thing to do was to carry the alternative-energy message to Los Angeles, Mumbai, and Shanghai.

I would do it with the *2041*. The boat had changed much since we first took charge of her. We had transformed the *2041* from a conventional racing yacht into a lean, green machine. The wind generators we had installed were now augmented by G24i solar panels fitted into the sails. We experimented with running the engine on various biofuels before settling on vegetable oil.

The Voyage for Cleaner Energy began in April 2008 as a four-year mission that will culminate at the next World Summit on Sustainable Development in 2012. My idea was to visit all five of the top carbon-emitting areas of the globe, beginning with the United States, continuing on to Europe and Russia, and finishing up in India and China.

My brother Charles, who lives in the United States, helped me immensely in understanding the mysteries of the American temperament. I considered Americans, and more particularly young Americans (cue David Bowie here) to be my customers, my target audience.

The bulk of the massive refitting of the *2041* was undertaken by the intrepid Ini. The *2041*'s core crew of Captain Mark Kocina and foredeck hand Jake Barrett were joined, along the course of the voyage, by student volunteers and guest sailors from our sponsors. It was a traveling road show that used waterways instead of highways.

Jake's story is typical of many 2041 workers. He was working in a bicycle shop in Reno, Nevada, when Annie Kershaw walked into his life. She had a bike repaired in his shop. The two made casual conversation. When Annie came back, she made Jake an offer that any sane man might have refused. "How would you like to go to Antarctica?" Annie asked him.

Jake blinked. "Sure," he said. "When?"

"Tonight," Annie said.

Jake had gotten his first passport only the week before. He had never been out of North America. But within twenty-four hours, he was swept into the 2041 maelstrom, journeying to E-Base, acting as Mr. Fix-It and then informal engineer on the 2041's Voyage for Cleaner Energy.

In the year 2041 people who are in college today will have matured into decision makers and power brokers. I sought to influence the future. I made sure the Voyage for Cleaner Energy focused on universities and colleges, with twenty-one schools scheduled for the first phase of the trip, the western United States. I thought of it in terms of investing in human capital. The payoff might be years down the road, but I hoped that when 2041 rolled around, the investment would reach maturity.

We launched the Voyage on April 8. Looking at my diary, I find it incredible that I was there at all. It was mad. Since leaving IAE and the E-Base on March 28, I had lectured in Hong Kong, carried the Olympic torch through St. Petersburg, and picked up Barney in Cairns. My head was spinning.

It was a sparkling, sunlit day on the Embarcadero in San Francisco Harbor.

Vivienne Cox, a vice president of BP, one of our major sponsors, represented the company at the launch. The initials used to stand for British Petroleum but now, with the company's push into alternative energy, stand for Beyond Petroleum. Vivienne, to whom I had been introduced by my old university chum Marcus Ware, really believes in the goal of an oil company transforming itself into an energy company.

San Francisco's director of Climate Change Initiatives, Wade Crowfoot, spoke, and Jared Blumenfeld from the city's Department of the Environment emceed the event. April 8, they informed me, was now officially Robert Swan Day in San Francisco. The event had a youth spin to it. Gavin Newsom, the youngest San Francisco mayor in over a century, is a go-ahead guy. (I can really appreciate a town that has its

own director of Climate Change Initiatives.) My own personal investment in human capital, the young Barney Swan, was at my side to witness the launch.

The whole Voyage crew embarked on a blitz of area schools. Stanford, Berkeley, City College of San Francisco, San Francisco State University, the Larkin Street Youth Center, Ocean Shores High School.

At every appearance, I tried to drive home a dual point. First, if you
have a dream, you *can* make it happen. Second, our collective dream of
a sustainable future *is* possible. I preached the gospel of reduce, reuse,
and recycle, but since I was fresh from my experience at E-Base, the
message had a little more punch to it. I had quite literally been there
and done that.

Whenever possible, I used the *2041* as a backdrop for my talk. She
was the world's greatest prop. I could point to the sails, made from recycled plastic bottles and fitted with solar panels. I talked about how
she smelled of French fries whenever we ran the biodiesel engines.

Was I reaching anyone? How many times can someone hear
"Change your lights to use energy-efficient bulbs," and "Properly in-
flate your tires" or "Ride a bike" before the mind shuts down? I didn't
really know. All I could do was present the message in the most force-
ful, immediate way possible, using our E-Base experiences and the
2041 to break through the mission fatigue of the audience.

On the voyage up the coast to Seattle, the crew got to test the *2041*
in fresh-gale conditions, with high seas and winds up to thirty-eight
knots. For much of the voyage they had "the bone in their teeth," to
use a sailing term for making good progress.

"She is no lady," Captain Mark observed. "When the waves come
at her she puts her shoulder down and drives right into them."

The Voyage was not without its random breakdowns and mishaps,
which I've wearily come to accept as somehow inevitable for any
Robert Swan project. On the whole, the biofuel engine worked admirably—as long as we were in the chilled, 60-degree F waters on the
American west coast.

But as we journeyed south to transit the Panama Canal and proceed
to the east coast phase of the tour, we passed through much warmer

waters. The seawater pumped in to cool the engine was nearer to 80 degrees F than 60. That made all the difference. The vegetable oil we were using overheated and gelled, turning into something that resembled black Jell-O.

The problem was that we didn't immediately discover the problem. The engine waited until we had actually entered "El Canal." Then it died, with the *2041* suddenly becalmed in the Miraflores Locks. We had huge cargo ships stacked up behind us as we tried to diagnose what had gone wrong.

We got the engine started again and we continued on to the Pedro Miguel Locks. It died again. By stops and starts we negotiated the whole of the canal. The operators were not sorry to see us go, once we exited into the Atlantic at Bahía Limón.

We were forced to switch back temporarily to diesel fuel. There were bumps on the road to the sustainable future, and sometimes it seemed like I was intent on hitting each and every one of them.

Nantucket

$\mathbf{M}$other Nature had the yacht *2041* on a tight schedule. We needed to get through the canal and north to Nantucket in time for an early-August event—a Barack Obama fund-raiser where I was to meet with Al Gore.

In an irony not lost on any of us, climate change slowed our canal passage. The Panama Canal operates on fresh water, the supply of which dwindled in the region as the global temperature rose. To be on time, we had to make it through the simmering waters of the Caribbean Sea in summer. One mistake we most especially did not want to make was to get caught in the Atlantic's "Hurricane Alley" in storm season. . . .

We got caught in the Atlantic's "Hurricane Alley" in storm season.

July 2008. Hurricane Bertha tore northward, up the east coast of the United States, dumping rain, following the course of the Gulf Stream as a category 3. Storms such as Bertha weren't supposed to form that far east, and they weren't supposed to last as long as Bertha did, seventeen days from its formation on July 3 until it petered out over the North Atlantic on July 20.

The next day Hurricane Cristobal picked up where Bertha left off, following a similar northeastern track. At the same time, Hurricane Dolly formed in the Caribbean, clipped the Yucatán on its way north-west, and made landfall near the Texas-Mexico border, knocking the stuffing out of Brownsville and Juarez.

At one point in July there were only four ships afloat in the area, dodging the B-C-D punches of Bertha, Cristobal, and Dolly. Three of those ships were oil tankers. The fourth was the *2041* under the steady hand of Captain Mark.

The temperature of the Atlantic in Hurricane Alley has grown warmer over the past decades, which climatologists believe increases both the size and number of hurricanes. The *2041*, a boat publicizing climate-change issues, was being stalked by climate change.

As she had proved on the Sydney Hobart Yacht Race, the *2041* was built for punishing seas. The Global Challenge, for which the *2041* was originally designed, is not for landlubbers. At 29,000 nautical miles, it calls itself "the world's toughest race." But even when her racing days were a memory and she had been transformed into a lean, green vegetable-oil-Jell-O machine, the *2041* still sailed fearlessly through what we were calling, with gallows humor on board, "the jaws of death."

I had to make it through, because in early August, I was invited to meet not only Gore but Bobby Kennedy, Jr., and the whole Kennedy family at their Hyannis Port compound. I couldn't let a little high wind keep me from a meet-and-greet with some of the leading lights of the American environmental movement. We scooted through the Caribbean and up the east coast to Massachusetts.

White, fluffy cumulus clouds rolled beautifully over Nantucket Sound, west winds came off the shore, and the temperature held steady at a brisk 60 degrees F. Bertha and Cristobal were long gone, in their death throes far to the east in the chilly waters of the North Atlantic.

Nantucket, the former energy center of America, the Houston of its day, had recovered quite nicely from its ghost-town status after the whaling industry collapsed. Tourism, not whale oil, was its lifeblood now. The streets of town came off as quaint and impossibly well heeled.

With the *2041,* I relearned the same lesson that had been beaten into me by the *Southern Quest*. A boat is a hole down which you pour money. I was facing financial collapse once again.

"We're going to have to pull the plug," Anne Kershaw told me, the voice of reason as my CEO. An end to the Voyage. No trip down the eastern seaboard, no more college appearances, mothballs for the yacht, a pulling-in and retrenchment for the whole 2041 organization.

It was with a financial sword of Damocles hanging over me that I found myself at the Obama fund-raiser, in a garden with Al Gore, Senator John Kerry, global-warming activist Laurie David, and a host of other heavy hitters on the American political scene. Powerful men and beautiful women, powerful women and beautiful men, all gathered in an impossibly gracious home of an impossibly gracious host.

My connections in Massachusetts were courtesy of an IAE 2008 alumnus named Sue LeCraw, who wore a stunning gown that evening and helped offset my sartorial sad-sackness. Sue had also introduced me to Laura Turner Seydel, CNN founder Ted Turner's daughter, who, along with her husband, Rutherford Seydel, I admire hugely for their environmental work, especially their program aimed at young people, Captain Planet.

As we assembled to hear Al Gore speak, he began by citing the multiple-crisis mode that had gripped the world, from the collapsing economy to numerous simultaneous wars and outbreaks of violence.

> The climate crisis, in particular, is getting a lot worse—much more quickly than predicted. Scientists with access to data from Navy submarines traversing underneath the North polar ice cap have warned that there is now a 75 percent chance that within five years the entire ice cap will completely disappear during the summer months. This will further increase the melting pressure on Greenland. According to experts, the Jakobshavn glacier, one of Greenland's largest, is moving at a faster rate than ever before, losing 20 million tons of ice every day, equivalent to the amount of water used every year by the residents of New York City.

Gore knew his stuff. He was preaching to the converted in that garden, for sure, but even so, I caught the same haunted look that I detected almost two decades before in the eyes of Jacques Cousteau. The look of people who had seen what was coming. It was as though Al had glanced into the mirror that morning and seen death's face staring back at him. His concern over climate change wasn't just political window dressing.

> I'm convinced that one reason we've seemed paralyzed in the face of these crises is our tendency to offer old solutions to each crisis separately—without taking the others into account. And these outdated proposals have not only

been ineffective—they almost always make the other crises even worse.

Yet when we look at all three of these seemingly intractable challenges at the same time, we can see the common thread running through them, deeply ironic in its simplicity: our dangerous over-reliance on carbon-based fuels is at the core of all three of these challenges—the economic, environmental and national security crises.

After his speech, I introduced myself to Gore, who had obviously been briefed on what I had been up to, since he had kind words to say about 2041.

"You've been seeing this kind of thing for twenty years," Gore said, referring to the degradation of the polar environment. "That's really important, to have that kind of longevity, so people know that it isn't just a recent development."

I told Gore that he had been instrumental in communicating inconvenient truths. "We're all about working on convenient solutions," I said.

"Fantastic," quoth Al Gore.

We both appreciated that we were in Nantucket, the onetime energy center of the world, a boomtown gone bust. "It happened before," I said. "The world found a new energy supply, the whales were saved from extinction, and now look at Nantucket." I gestured around at the beautiful people in the beautiful garden. "We can do it again," I said. "We can find new energy sources and turn it around."

Gore looked unconvinced. He flashed the weary look of an old warrior who feared we were losing the war. "You know, I haven't seen one wind turbine or solar panel since I've been here," Gore said.

"Nothing," I agreed.

He sighed. I sighed. He looked around at the glittering crowd. "And it's not like the people here can't afford it."

The United States puts over six billion metric tons of carbon dioxide into the atmosphere per year, almost a quarter of the world's total.

Sensible national standards for automobile emissions and mileage could cut that to four billion metric tons per year within a decade.

Had Al Gore taken office in January 2001, one of his first acts would have been to tighten emission and mileage standards. George Bush did nothing but encourage the American auto industry's foot dragging. Over the course of the eight years of his presidency, sixteen billion metric tons of additional hydrocarbons were loosed into the atmosphere that could have been eliminated by simple presidential fiat.

The imminent tipping point on climate change is such that we can't afford to lose a single year, much less eight.

The next day, the beautiful weather disappeared. A cold, drizzling rain set in. I was disappointed. We were scheduled to meet the Kennedy family at the home of Chris Kennedy, one of Robert Kennedy's sons and a mover and shaker in Chicago business and politics. I felt sure none of the Kennedys would come out in such poor weather and that my day would be a bust.

We docked the yacht at Hyannis Port and I set off alone in search of a taxi to bring me somewhere where I could rent a car. As I left the marina a car pulled up, its back window rolled down, and I was looking at the familiar and distinctive Kennedy face.

"You must be Rob Swan," Robert F. Kennedy, Jr., said. "What are you doing walking about in the rain?"

I told him I was looking for a cab, and he told me to hop in, that he'd give me a ride to my rental car.

Robert F. Kennedy, Jr.'s work with water conservation and river cleanup deserves the Nobel Peace Prize. Soon enough water wars are going to join energy wars as the bane of the planet, unless we have more committed people such as him and his wonderful wife, Mary Richardson Kennedy.

It turned out that the poor weather actually worked in my favor. Gathered at Chris Kennedy's home that day were not just a few Kennedys. It was the whole clan, almost one hundred people, distinguished elders such as Bobby's mother, Ethel; notables such as Maria Shriver and her mother, Eunice Kennedy Shriver; plus dozens of younger children. On a nicer day most of these people would be out windsurfing on Nantucket Sound, or sailing to Provincetown—doing something, anything but listen to a boring Brit give a bloody talk and slide show.

I was there to tell them about climate change and Antarctica. I admit to taking a deep breath before getting up in front of that crowd, with Ethel in the front row along with Maria, Bobby, and Mary. The Kennedys were a family of orators. They had heard some of the most stirring speeches of the modern age. I was quite sure I would not measure up.

But I launched into it anyway. "Laura Seydel was kind enough to introduce me as the first person to walk to both the North and South poles," I began. "I'd like to amend that sentence. I'm really the first person *stupid enough* to walk to both poles."

Laughter. I had successfully broken the ice, so to speak. I had them, a hundred members of one of the most powerful and activist families in America. I was especially gratified by the rapt attention on the part of the younger members of the clan. They were my true audience. They would be the ones making decisions in the year 2041.

Bobby and Mary Kennedy were all about clean-water initiatives. After working with the Hudson's Riverkeeper group, Bobby helped sue polluters, clean up the Hudson River, and protect the watersheds that kept New York City drinking fresh tap water. Bobby and Mary ran an umbrella organization called the Waterkeeper Alliance that concentrated on water issues around the globe.

When I finished my talk, Bobby got up in front of his family. "After Rob reminded me that seventy percent of the world's fresh water is locked up in Antarctic ice, Mary and I thought it might be fitting to give him a new title to go along with all the other awards and accolades he's gotten over the years."

I had no clue what he was talking about. I hoped for something along the lines of being made an honorary Kennedy. But Mary and he went one better: On behalf of the Waterkeepers, they officially named me the honorary "Icekeeper."

Robert Swan, Icekeeper. I liked that.

Hearing Gore and then Bobby Kennedy speak electrified me. They both questioned whether the 2041 idea was, in fact, too generous. They spoke more in terms of 2020.

I called Annie from the satellite phone on the boat. I didn't care if I had to spend my own money, I said; we were going on with the Voyage on the *2041*.

I heard an audible sigh from the other end of the line. She had heard this story many times before. She was weary of trying to save me from myself. She got on the telephone to my great friend and generous supporter Simon Freakley of the firm Zolfo-Cooper, who would help us yet again dodge bankruptcy. Anne called me with news of a patch that would get us through the crisis.

Onward, I told her.

Why Antarctica?

I persevered. The Voyage for Cleaner Energy continued. As of this writing, the *2041* is heading to Europe and Russia, then on to India and China, to wind up eventually at the 2012 Earth Summit.

I encountered the tried-and-true questions whenever I spoke to audiences on the Voyage. "Was it cold?" they ask of my walks to the North and South poles. The second most popular query: "How do you go to the bathroom when you're on a polar expedition?" I usually dispose of these queries the same single-word way, with "Yes" and "Quickly."

But the most important question, one that I often got also, deserves a more involved answer. It took many forms but amounted to the same thing. What makes Antarctica important? Why should we care? Why is this particular continent the global version of a canary in a coal mine?

There were several ways to respond. I could talk about the beauty of the Antarctic skies, the impossible colors of ice, the wrecked majesty of a glacier. Or I could articulate Peter Scott's plaintive plea to leave just one place on earth alone.

On my travels and during my talks, I've found that this way of answering the question leaves some people cold, if you'll pardon the pun. The "great beauty" argument and the "untouched wilderness" argument depend to some extent on subjective judgments. The only way to reach the obdurate few in my audiences is with hard facts. The most effective argument always boils down to an appeal to self-interest. We were never more popular than when the price of gas rose beyond $4 per gallon.

The environmental movement as a whole has found this to be true. All the slogans about harmful emissions and carbon footprints had less impact on the popularity of SUVs than did a few months of four-dollar-a-gallon gas. The most persuasive line will always be the "green equals green" argument, the idea that being environmentally aware can mean money in the pocket.

In this sense, the foundational argument for preserving Antarctica is based on hard science. The latest climate models can work to convince even a die-hard skeptic, someone who might remain unmoved by the great beauty or majestic isolation of the continent. Thanks to the latest research, we can explain cogently why Antarctica serves as a reliable canary in a coal mine for the rest of the planet.

A stone in the shoe of climate-change activists, and one of the favorite counterarguments of climate-change skeptics, is the widely circulated claim that Antarctica is actually cooling. Like so many aspects of the debate, the cooling of Antarctica is not at all as clear-cut as is usually claimed.

Parts of Antarctica may be cooling. As climatologist Susan Solomon (the same woman who did such pioneering work about the weather during Scott's expedition) has indicated, research shows that because of complex atmospheric and chemical interactions, the destruction of the ozone contributed to a cooling trend for a specific area of the continent. It is a trend not likely to be maintained.

To understand what is happening to Antarctica's climate, it helps immensely to see it not as a uniform, homogenous place but rather as a landmass with several different areas. When it is considered like this, the science and research become clear: The overall trend in Antarctica is warming, just as it is for the rest of globe.

Antarctica breaks geographically into two large sections and one smaller one. The two larger are usually labeled East and West Antarctica, though some prefer Lesser and Greater Antarctica. The third area is the Antarctic Peninsula, the "hitchhiker's thumb," pointing toward South America. Dividing the east and west, the greater from the lesser, are the ranges of the Transantarctic Mountains, which bisect the continent in a line running from the Ross Sea to the Weddell.

East Antarctica represents much the larger chunk, roughly three-quarters of the whole, a vast, empty expanse embracing the high, dry, and windy Antarctic Plateau. West Antarctica and its smaller adjunct, the Antarctic Peninsula, is the most explored and populated part of the continent.

Both East and West Antarctica are covered by massive caps of ice, called, conveniently enough, the East Antarctic Ice Sheet, or EAIS (pronounced to rhyme with "peace") and the West Antarctic Ice Sheet, or WAIS (rhymes with "ace").

The EAIS is one of the most incredible topographical features in the world. It is more than ten thousand feet thick in some places and as large as the continental United States; its huge mass depresses the ground beneath it by as much as a mile.

Glaciologists recently extracted an ice-core sample from Dome C of the EAIS that traced 800,000 years of climate change—back to the Pleistocene, before *Homo sapiens*, when *Homo erectus* was just taming fire. It turns out that the EAIS is fairly stable in terms of climate, protected from influxes of warmer air by the Transantarctic Mountains and a steep-sided barrier rim. This is a good thing, since the EAIS locks up an immense amount of water in ice. If it ever melted, sea levels around the world would rise two hundred feet.

The EAIS is also stable in terms of total mass. The amount of snow falling on it roughly balances out its loss from glacial action (the mighty Beardmore draws off the EAIS). In terms of climate change, scientists like the stable, balanced nature of the EAIS.

The WAIS and the Antarctic Peninsula worry climatologists much more. This is largely because the WAIS is not land-based like its sister ice sheet to the east but marine-based—meaning it meets the ocean and its underside is actually below sea level.

The collapse of the WAIS is a climatologist's version of the apocalypse. While it is not imminent, the fact that it is possible scares a lot of very serious-minded people. The WAIS locks up only 20 percent of Antarctica's water, but even so, a collapse could raise worldwide sea levels by twenty feet.

If Antarctica is the planet's canary, then the Antarctic Peninsula is the canary for the continent, a canary's canary. Compared with the WAIS, it is smaller and much more exposed to climate influence from the seas that surround it. What we have seen since 2000 is a rapid disintegration of the glaciers and ice caps in this sector of the Antarctic.

In 2002, the canary's canary staggered and swooned. The Larsen B Ice Shelf collapsed. Larsen B was a feature of the east coast of the peninsula that (studies would show) had essentially remained intact for ten thousand years. A portion the size of Rhode Island broke off and disintegrated.

The Larsen B was the ninth such Antarctic ice shelf to disintegrate and only gained so much notoriety because it was by far the largest. In 2009, the peninsula's Wilkins Ice Shelf is in its "death throes," according to a BAS scientist, hanging on by a thousand-foot thread of ice.

In terms of temperature change, glacier retreat, ice breakup, and snowmelt, the Antarctic Peninsula has been changing at an accelerated rate. Temperature records indicate an increase of five degrees since 1950 on the peninsula, and there are strong indications of a three-and-a-half-degree increase over the same period on the WAIS. We should take heed.

The image I get from these research studies is of climate change hitting the Antarctic Peninsula hard and marching south toward the larger mass of the WAIS. Computer climate models project that, due to greenhouse-gas effects in higher latitudes, the circular band of winds surrounding the continent will produce bigger and stronger storms.

This in turn will send more cyclonic systems of warmer air across the exposed peninsula and into the interior of West Antarctica. More melt, more glacier retreat, more ice breakup.

This is not "theory." This is not "junk science." The change of climate in the Antarctic Peninsula has been conclusively demonstrated in hard, crystal-clear numbers. The canary has not fallen off its perch just yet, but it is clearly wobbling and staggering.

If scientific research remains unconvincing, I offer my own anecdotal observations. In the twenty years I have been going to Antarctica, since my first visit to the peninsula with the BAS at Rothera Station, I have witnessed change myself. I have noticed the tongues of glaciers retreating from the sea, bigger and more numerous icebergs, many of them recently calved, and warmer temperatures all around.

We have to make change happen in temperate and tropical zones where most of the human population is concentrated. If we don't, then the changes already evident in the polar regions will move out of the Antarctic and the Arctic and affect us where we live.

"Why Antarctica?" has another answer, too, perhaps a more sobering one. Why preserve Antarctica? Because it's already too late for the Arctic.

The view of the North Pole from the South Pole seems increasingly fraught. In August 2007, a Russian flag planted on the ocean floor two miles beneath the North Pole indicated a creeping nationalism that could make the Arctic and Antarctic pawns in an international game, places for confrontation rather than cooperation.

No one owns the South Pole, and no one is supposed to own the North Pole, either. It's controlled by a U.N. commission, with the territorial claims of the eight polar countries—Russia, Sweden, Finland, Iceland, the United States, Canada, Norway, and Denmark—limited only to the traditional two-hundred-mile strip along each nation's coastline.

The 2007 Russian land rush—the *New York Times* labeled it "An Arctic Game of Chicken"—traced the underwater Lomonosov Ridge from the coast to the pole, thus providing a lebensraum logic for the flag on the ocean's floor.

Under Clause 76 of the International Law of the Sea, a geographical "prolongation" of a country's continental shelf can lead to territorial claims beyond the generally recognized coastal boundaries.

The Russian Lomonosov Ridge rationale was rejected by the U.N. Commission on the Limits of the Continental Shelf. But Russia does not seem to be ready to back down, and continues to push its claims.

Russia isn't alone. China now has a polar-research station on the Norwegian island of Spitsbergen. The country has sent one of its Antarctic icebreakers, the *Snow Dragon*, north to the Arctic. Denmark, Canada, and the United States have all mounted mapping projects for the floor of the Arctic Ocean. Lloyd's of London reports an abrupt spike in the number of ice-hardened ships registered to different countries around the world.

The Arctic is heating up, literally and figuratively. Of the world's yet-undiscovered oil and gas resources, a full quarter is located in the Arctic. Already, the pressure to open more of the Arctic to exploration and exploitation is increasing.

By 2041, the fabled Northwest Passage, the goal of so many expeditions and so much exploration, may finally become an ice-free reality. Freighters and supertankers will be able to travel the "Arctic bridge," from the North Sea oil fields and the refineries of Norway straight over the top of the world to the thirsty oil markets of Asia. The new shipping lanes are thousands of miles shorter than traditional routes because of melting ice.

Looking to the north from Antarctica, could we be seeing a version of the future? Will the Big Melt come to the south, too, and trigger a subsequent land rush?

That flag on the floor of the ocean beneath the North Pole reminds me of those small metal swastikas that the Nazis dropped to establish their territorial claim to Antarctica. The thought behind that flag is just that ugly. I've personally seen some strange things in my life, but one of the strangest is what's currently happening in the Arctic. We've melted the ice, which means it's now possible to bring in the machinery of petroleum exploration and exploitation. Why did the ice melt? Because we rely on fossil fuels for energy. Why are oil men taking advantage of the melted ice to scour the Arctic? To search for more fossil fuels. We need to connect the dots.

Antarctica advocates can only look at the Arctic in dismay, rededicating ourselves to never allow history to repeat itself, and to prevent what is already happening in the north from occurring in the south, too.

The Return from the South Pole

As I write this, the fourth International Polar Year draws to a close in March 2009. This is also the fiftieth anniversary of the signing of the Antarctica treaty in 1959. Three previous International Polar Years have been held, in 1882–83, 1932–33, and 1957–58. The sobering truth is that this may be the last one. Unless present rates of climate change are halted and eventually reversed, the next International Polar Year may be a gathering of historians.

I am working at my makeshift desk at E-Base. The space is heated by renewable energy, and my computer is powered by the same. I can communicate to the world from here without resorting to a teaspoonful of fossil fuel.

Writing from E-Base naturally puts me in a philosophical mood. Directly in front of me is the beach at Bellingshausen and beyond it, Collins Harbor. To the right I can see the assembly of industrial-style buildings that is the Russian base. At my back, through a high window, I can view the desolation of the Valley of the Shadow.

Headed by the irrepressible Kevin McCullough, the team from RWE N-Power/Inogy is here again, with new wind turbines, refining our approach to renewable wind power. When Germany's largest electricity provider, under the visionary leadership of Jurgen Grossman, commits itself to an alternative-energy project like E-Base, I take it as a very hopeful sign.

In the 1930s, in the face of the Great Depression, President Franklin Roosevelt announced the initiatives of the New Deal. Today we need a Green Deal, an initiative just as dramatic, just as comprehensive, just as far-reaching, designed to wean the industrial world off the fossil-fuel teat.

As Al Gore indicated in his speech at Nantucket, this will have a salubrious effect on a whole range of problems, seemingly discrete but really connected—the economy, the environment, national security.

When President Kennedy announced his goal to put a man on the moon, the young people in his audience were in their teens. Less than a decade later, when Neil Armstrong and Buzz Aldrin landed on the Sea of Tranquility, the average age of the engineers and specialists in the control room at Houston was twenty-six. That's how the torch of the generations is passed. One generation inspires another with a great task. The fire of youthful idealism generates the heat to get it done.

We're setting the mission to preserve Antarctica. How can we accomplish this? By championing renewable energy, the very stuff I am using to write these words at E-Base, and promoting its use all over the globe. As President Kennedy's moon-shot challenge resulted, a decade later, in all those twenty-six-year-olds working in a Houston control center, our challenge has to be addressed to the young. Who we're after now are eighteen-year-olds. To preserve Antarctica, we have to get out there and inspire young people.

I'm not stopping. I continue to communicate the idea of 2041 to the young people of today. Every year we take a full component of leaders—both students and established business professionals—on 2041's expeditions to Antarctica for them to see for themselves.

Thanks to BP, I am in Antarctica at E-Base with the seventh IAE, and seventy team members. Amazingly, due to the efforts of Vivienne Cox and the rest of the BP Alternative Energy team, we'll also be welcoming seventy students to become the next generation of Antarctic champions. In the worst recession in decades, we've still managed to demonstrate our commitment, with funds, time, and energy, to the cause of preserving the polar wilderness. Is that not hopeful?

On a personal level, the future also looks promising. I have once again decided to heed the call of the "wild corners of this dreadfully civilized world."

For a long time, I have resisted that persistent call. I didn't want to be cold and miserable again, facing the desolate challenge of the Antarctic landscape. But always, a phrase recurred in my mind. It's from the notebooks, but I first heard it spoken by Johnny Mills in the movie *Scott of the Antarctic:* "Now for the run home, and the desperate struggle that lies ahead."

The doomed trek back from the pole by Scott, Wilson, Bowers, Oates, and Evans continues to haunt me. It is still, for me, a potent mix of tragedy and heroism.

I had the nagging sense of unfinished business. I didn't really want to embark on another "desperate struggle." But I had aborted my participation in the One Step Beyond Expedition. And that damned phrase from Scott continued to dog me. "Now for the run home . . ." Part of me will be dragged into it kicking and screaming, but it seems there is nothing for me but to mount another Antarctic expedition.

In the South Pole summer of 2011–2012, with a team of five expedition mates, I will commemorate the one hundredth anniversary of Scott's South Pole expedition with a trek tracing his return route. The Return from the Pole Expedition will not only be a tribute to the past but also an effort to grapple with the technical questions of the future.

The survival of human life on earth in this new millennium has now become a technical challenge, revolving around how we produce, use, and save energy. Apart from our survival on the planet as a species, that technical challenge directly affects the last great wilderness on earth. If we find a way to utilize more renewable energy in the real (temperate zone) world, then it will make no financial sense to exploit Antarctica.

In other words, unless we fail to employ new forms of energy and the price of oil heads upward past two or three hundred dollars a barrel, there will be no reason to extract the petroleum deposits at the poles. Some firm of accountants (such as my old friends at KPMG) will crunch the numbers and, with a click of the computer keyboard, will thereby save Antarctica.

"Sorry, blokes," the accountant will say to the oil wildcatters, "it's simply not worth doing." *That's* how Antarctica will be preserved as wilderness.

My own personal response to the technical challenge of the future, and the response of the 2041 organization and whichever friends I can cajole into doing it, is to undertake the first-ever polar expedition powered solely by renewable energy. Nine hundred miles to Cape Evans, back from the pole, reversing the route Footsteps took twenty-five years earlier, all the while using not one tablespoon of fossil fuels. No white gas, no diesel, not even paraffin.

As I write these words, the technology does not yet exist to make such a journey in safety, so we have to develop new methods ourselves. We are currently on the task, and I know we can find a way to succeed. Our search for solutions is a microcosm of the challenges that the whole world faces today.

Barney Swan and a youth team will meet the Return from the Pole Expedition out on the Great Ice Barrier, near where the body of Captain Scott still lies. Together, we will complete the journey to Cape Evans. I and my grizzled expedition mates will symbolically transfer to a new generation the challenge of preserving Antarctica. It will be time to hand over the baton.

While hiking at Ayers Rock in Australia, that mystical center of the universe perfect for life-changing decisions and new resolutions, my son and I made a pact to complete the last portions of Scott's tragically aborted expedition. This will be much more than a simple father-son outing. In December 2011, Barney will be seventeen, almost eighteen. He will symbolize the rise of new leaders, new advocates for Antarctica. If his generation takes up the challenge, we have some hope of preserving the earth's last great wilderness.

When giving me the fifty-year mission to preserve Antarctica, my mentors Sir Peter Scott and Jacques Cousteau both recommended a similar strategy. Develop short-term missions, they said, that people can understand and get behind. Inspire people in the short term. The Return from the Pole expedition is one of those short-term missions that lead to the larger goal of preserving Antarctica.

"You'll be surprised how quickly the long-term mission passes," Sir Peter Scott said to me. "We live on this earth as visitors. All human missions are necessarily short term. But what are we here for, if not to pursue them?"

As I write this, I am eighteen years into my mission, and thirty-two years from 2041. How right Peter Scott was.

Transforming the future but not letting go of the past. I think about my mother's life. Em was born in 1915, when automobiles were rare.

When the family did get a car and she was old enough to drive, there were no licenses, no road tests. She just got in and drove. Such procedures were more informal then, even if social interactions were much more formal. My mother's generation has seen transformations that are awe-inspiring—radios, space travel, the Internet. It is hard to grasp the fact that Em was born only two years after Scott died. Her life encompasses almost the entire period of human impact on Antarctica.

Scott's life, my mother's life, my own life, and my son's—over the course of only four generations, really, we will determine the future of the world's most ravishing wilderness. In 2041, Barney will be in his mid-forties, in the prime of his life. The die will be cast for Antarctica in that watershed year.

The decisions of the past and the present extend inexorably into the future. In the 1990s, the measurements of a team of American researchers led to a macabre prediction. The ice of Antarctica moves not only at the South Pole but all over the continent. That includes the spot on the Ross Ice Shelf where Scott, Wilson, and Birdie Bowers died of exposure and starvation on the return from their trek to the pole.

In a fitting memorial, while their diaries and personal effects were removed, the mortal remains of the three were left in their tent after its discovery by fellow expedition members. Gradually, snow covered it over, and the ice pack entombed the bodies.

But the restless ice continued to move. The body of Captain Scott and those of his two comrades migrated about a thousand yards a year, according to the American research team's calculations. Their corpses move north along with the rest of the Ross Ice Shelf, grinding steadily toward a terminus in the Southern Ocean.

The end of Captain Scott's journey would finally come, the Americans said, decades into the future. The portion of ice entombing Scott's remains will split off from the ice shelf as an iceberg, drop into the sea, and drift north to melt in temperate waters.

That would happen, the American research team predicted, sometime in the year 2041.

Macabre as it is, the idea of Scott's final journey naturally turns my mind toward the philosophical. I think of myself standing on the Great

Ice Barrier, or at the pole somewhere above Amundsen's black tent. I wonder where my own journey will end, and where the human journey itself might end. Yes, it has already been a long, strange trip, and no, it's not over.

Yes, there are any number of hopeful signs. In the past quarter century, we've made positive strides on the environment through concerted cooperative effort. One example? That hole in the ozone layer, the one that turned my blue eyes gray back in 1985, has been contained seasonally and is gradually closing and getting smaller.

The improvement was primarily due to the Montreal Protocol, a treaty signed the year after In the Footsteps of Scott. Montreal mandated aggressive regulations regarding Freon and other industrial byproducts. We changed our ways in the service of communal well-being.

How many people are aware that the hole in the ozone has been largely stabilized? I am betting not as many as are fretting over the cataclysm of climate change. It's good news, and good news sometimes doesn't travel well in environmental circles. If I got caught out in Central Park in a drug sting, say, the scandal would probably have much wider currency than anything else I've ever done, including the pole walks and the Bellingshausen cleanup. But I am not looking for that kind of notoriety. I am hoping to be caught out doing the right thing. To preserve Antarctica, we've got to be in the good-news business, and get word out of positive developments such as the stabilization of atmospheric ozone depletion.

It can be done, and it's important to remember and embrace the fact that it has been done.

The sole reason for that flag on the ocean floor at the North Pole was a vision of an ugly future where humankind fights over the last scraps of the planet's energy resources. If this vision of the future comes true, it doesn't really matter whether the polar environments are destroyed. In 2041, if we are that desperate for oil, the planet itself will have gone apocalyptic. It will have much wider implications than the rape of the last wilderness. It will mean that humankind has not solved the problems of energy supply and climate change. And that in turn will mean the end of a human-friendly planet.

Here at E-Base, I feel hopeful. We represent a tiny colony of the future, a lone outpost in a harsh environment, erected in horrible conditions, powered entirely by renewable energy, and connected to the whole globe, both literally, via BGAN, and figuratively, through the idea that everyone on the planet is linked to everyone else.

The wind turbines perform better this year. The team from RWE N-power/Inogy has been fantastic. E-Base has now gone live year-round, even when no one is in residence, so that schoolchildren can always check in electronically and see what's going on at the bottom of the world. In the long Antarctic night, we'll be only a small pinprick of light, but a beacon, I hope, pointing the way to 2041 and beyond.

What You Can Do

Until one is committed, there is hesitancy, the chance to draw back, always ineffectiveness. Concerning all acts of initiative (and creation), there is one elementary truth the ignorance of which kills countless ideas and splendid plans: that the moment one definitely commits oneself, then providence moves too. A whole stream of events issues from the decision, raising in one's favor all manner of unforeseen incidents, meetings and material assistance, which no man could have dreamt would have come his way. I learned a deep respect for one of Goethe's couplets:

Whatever you can do or dream you can, begin it.
Boldness has genius, power and magic in it!
—W. H. Murray,
The Scottish Himalayan Expedition

Life feels a bit lonely lately. We've lost the inimitable voice of Jacques Cousteau. From the U.K. alone Sir Peter Scott, Sir Vivan Fuchs, Lord Hunt, and Lord Shackleton have gone. I hope the vanguard of today's environmental movement, Jonathon Porritt chief among them, will agree with my humble suggestions that follow.

The most potent threat to addressing climate change is the conviction that someone else is going to do it. What can an individual do? What can you do?

Well, there's one option that we all have: Learn how to swim. If the West Antarctic Ice Sheet continues to degrade, there will be a lot more opportunities for aquatic locomotion.

Learn how to swim, learn how to sweat, learn how to put up with food shortages and population dislocations and the myriad other changes coming our way, as long as we continue to cling to the idea that someone else will somehow, in some way, reverse climate change for us.

But say I've convinced you, over the course of reading this book, that Antarctica is worthy of preservation. Or perhaps you've always harbored that belief. But believing that something must be done is different from actually helping to do it.

I would like to be able to tell you that this is an exhilarating crusade, that you must buy yourself a big white steed and gallop across the frozen Southern Ocean bearing a long streaming banner that reads SAVE ANTARCTICA! You'll be hailed as a savior.

But the real path before us is a great deal less sexy than that image. So the first step you can take to help Antarctica is to give up dreams of glory in order to grapple with the mundane.

The day-to-day is boring, the day-to-day is dull, but multiplied by hundreds, thousands, millions, and finally billions of people, the day-to-day makes change happen. Not exactly the way of the white steed but rather the good old-fashioned nuts-and-bolts approach.

The single measure I ask of every person who travels to Antarctica with our annual IAE groups is simple. *Calculate your carbon footprint.* There are many methods to determine this figure, including some good ones that are online. When you know how many tons of greenhouse gases your household is generating, you can begin to address the issue.

Oddly enough, the fixes that generate the biggest bang for the buck are the ones nearest home, the energy-saving lightbulbs, the additional insulation, the fuel-efficient car. If you look at the bang-for-the-buck cost-efficiency rankings for all sorts of measures designed to respond to the climate-change crisis, you will see that solar, geothermal, wind, and other renewable-energy sources barely dent the chart.

What's the one thing we can do that provides the most impact for the least money? Insulate our homes. Solar panels partake a little more of the white-steed approach, good but not best for an initial step. Embrace the mundane. Insulate first, investigate alternative-energy sources after.

Next, direct yourself to some political nuts and bolts. In Copenhagen in 2009, and at the next Earth Summit in 2012, gatherings of world leaders, politicians, and decision makers will help determine the future of our world. It is vitally important that we exert every bit of our influence as voters, advocates, and common, ordinary earth dwellers to affect the agendas of these summits.

Since it will largely determine the program of the 2012 Earth Summit, the December 2009 U.N. Climate Change Conference in Copenhagen is the more immediately important. In 2012, the Kyoto Protocol on greenhouse gases is set to expire. Copenhagen will lay the groundwork for a new, urgently needed, global climate accord. Governmental representatives from 170 countries will attend. They'll be met by members of advocacy groups, NGOs, journalists, and others. Ten thousand people will descend on Copenhagen for the climate meeting, hosted by Prime Minister Anders Fogh Rasmussen and Connie Hedegaard, the Danish minister of climate and energy.

This conference must produce more than, well, hot air. Because Copenhagen establishes the terms of discussion for the future, I believe it is one of the most important gatherings in history, right up there with the meadow at Runnymede in 1215 or the Conference on International Organization at San Francisco in 1945.

That sounds like hyperbole, but it isn't. The Magna Carta and the United Nations will both become more tenuous if the degradations of climate change hit the globe full force. The effects of a five-degree mean-temperature change can turn economic and political challenges into biblical-scale catastrophes: war, pestilence, the storm winds of Job.

It has been said that democracy is a product of cheap energy. If those who hate democratic institutions wish to demolish them, they need to do nothing more than create a machine that changes the atmosphere of the earth, destroying the peace and prosperity in which democracies thrive. But wait—that machine has already been created. It's called the internal combustion engine. So Copenhagen is vital to address not only the environmental repercussions of climate change but the political ones, also. It's a peace conference. It's a meeting that will impact global poverty. It's all of that, all rolled into one.

We'll be in Copenhagen with the *2041*. And we'll take her to Rio in 2012 for the third world summit.

In addition to lending our voices in advocacy for the Copenhagen conference, and to such mundane tasks as inflating our car tires to the correct pressure, winterizing our homes, and changing our lightbulbs, there's a third step you can take toward preserving Antarctica.

Go there.

This step is not realistic for everyone, and it comes equipped with plenty of caveats and quid pro quos, but the scales tip toward my recommending it. A visit on a properly run polar expedition has much less impact on the fragile Antarctic environment than do the permanent scientific bases scattered around the continent. And it has the almost-inevitable effect of galvanizing visitors into becoming Antarctic advocates.

Traveling the length of the Gerlache Strait and the Lemaire Channel, with the snow-streaked teeth of the mountains of the Peninsula and Booth Island raised above their thick mantles of ice, the visitor feels the majesty of nature press upon him as in few other places.

For me, many of those other places are in Antarctica, also: the Great Ice Barrier, the McMurdo dry valleys, the Beardmore. The sight of a gentoo penguin waddling toward me might trigger thoughts on the stubborn persistence of life in the harshest environments. Looking into the blank-eyed stare of a leopard seal leads me to confront my place in the chain of being. The desolation of the Valley of the Shadow at Bellingshausen helps burn clean my soul.

I have traveled to Antarctica with hundreds of people. I have never seen one of them remain unstirred at the beauty of the place or unmoved by its challenge as the last pristine wilderness on earth. This is partially because it takes such a huge commitment to get to the continent that none but the brave and strong attempt it. The winnowing process is just that severe.

A trip to Antarctica is self-selecting. You will meet some of the most interesting, bright, passionately involved people on the planet. They will inspire you, and you will inspire them. The whole of the group will equal more than the sum of the parts. You will come away refreshed, renewed, and inspired.

The cynic will note that every visitor to Antarctica causes an immense load of carbon to be released into the atmosphere. I am well aware of this, and yet I still recommend people make the trip. We've done the math. Each person who visits Antarctica generates ten tons of greenhouse gases just to get there and back, including the flights to and from South America. That's the equivalent of the annual emissions output of an extra-large (the auto industry calls them "full-size") SUV.

Purchasing carbon offsets can help alleviate the impact. But what if we divide those ten tons by the number of years until 2041? Then the task becomes clearer. Then visitors to Antarctica must offset their initial carbon emission to the tune of around 0.625 tons per year, every year from now until 2041. That sounds more doable. The benefits of an Antarctica visit, through advocacy and direct action, must outweigh its environmental impact. Each visitor must leave the continent with a mission to protect it. Their missions must counteract their emissions, so to speak. It's now possible to make a carbon-free virtual visit to Antarctica via today's amazing online technological tools. You may visit it virtually or in person, but either way, go there.

So, these are three tasks I lay before you. Act on the home front, in your everyday life ("clean your rice bowl" as the Zen master says). This includes knowing your carbon footprint. Then raise your voice for the Copenhagen and Rio world gatherings. And last, if you possibly can, travel to the bottom of the world.

All of which brings us back to E-Base, and to 2041.

Thirty-two years from the time of this writing, in the year 2041, the polar regions will be able to tell us if we are saving the planet for human life or destroying it. If in 2041 global warming has continued unabated, the polar regions will be hardest hit, and we will have failed. If in 2041 the international community cannot summon the cooperation necessary to preserve Antarctica as "a Natural Reserve Land for Science and Peace," then we will have failed.

Antarctica has seen the best, most intrepid, and certainly most courageous acts that humankind can offer. It has also seen some of the worst. Both sides of the human personality have left their marks on the clean slate of the continent. From where I stand, the two sides always exist in opposition, between preservation and exploitation, knowledge and ignorance, education and self-interest.

No oil has yet been discovered in Antarctica, but strategic ores and minerals have been found. The climb of commodity prices for such staples of industrialization as bauxite, copper, zinc, and magnesium will perhaps turn profit-minded eyes south before too long.

We've largely held out against such pressures so far. What I like about international cooperation in Antarctica is what it says about us as human beings. It means we can make a choice to honor what is right over short-term profitability. If we can continue to do this in Antarctica, perhaps we can apply the same strategy in the world as a whole and turn back the cataclysm of global warming.

I hope that Antarctica will be ruled by what Abraham Lincoln called "the better angels of our nature." That's been my ideal ever since that faltering promise out on the barrier—help people to discover that this land that I love is a place that's ultimately worth preserving.

The challenge of modern life is not a lack of information or even of resources. It's a lack of inspiration. My job, and the mission of the 2041 organization, has been to inspire people, to get them to believe that a journey is possible by showing them small, achievable steps.

My lowest ebbs, the loneliest, most desolate, and dangerous points in my life, weren't always on the ice. At times there were battles to be fought in my personal life, too. But I've always clung to the singular truth that the power of positive action can change everything.

I want to apply my ideas of leadership to the goal of battling human-induced climate change. We face the same problems in grappling with the issue that we always face in any team-building or goal-oriented endeavor: apathy, divisiveness, lack of leadership.

What do we need to fight human-induced climate change? Far-thinking international cooperation is essential. Where is this kind of cooperation most effectively embodied today? In the Antarctic Treaty System.

That means that Antarctica helps us to measure not only the problem of human-induced climate change but also the possibilities of humanity working together toward a solution. The challenge now is to inspire the young people of today to recognize the importance of 2041 as a goal, a test, and a concept.

I dedicated this book to my mother and my son, who are eighty years apart in age. I think of all the enormous changes that have happened in Em's lifetime, not only in the world in general but in Antarctica in particular. What will the continent look like when my son is my mother's age?

Barney has just entered his teens. He loves skateboarding and mountain biking. He brings me balance.

"Hello, Dad?" he says, pulling me up short when I embark upon a lecture. "Don't always be finding solutions when I have a problem."

The two sons of great Antarctic explorers whom I have met—Sir Peter Scott and Lord Shackleton—both voiced the same sentiment in my conversations with them, using almost identical phrasing. "It's not easy being the son."

As far as I can tell, Barney is weathering the storm well. He has committed himself to being a doctor. At age twelve, he went off on his own accord and booked himself into a first-aid course, the youngest in the class by a good fifteen years.

The stories that I tell Barney, about In the Footsteps of Scott and Icewalk and loveLife and E-Base, all have a common theme. They're about what is truly possible in this world. They're not just my stories. They're our story. They're for Barney and for everyone in his generation.

In the time it takes you to read this book, around .06 percent of the time between now and January 1, 2041, will have elapsed.

We had better get cracking.

November 2007–March 2009
Wood Wharf, London
Sofitel, New York City
E-Base, Bellingshausen, Antarctica

Appendix 1

Robert Swan and 2041 Time Line

1979	B.A. Honors, Ancient History, Durham Univeristy, United Kingdom
1980–81	Field worker, British Antarctic Survey, Rothera Research Station, Antarctica
1984–87	In the Footsteps of Scott South Pole expedition; South Pole reached January 11, 1986
1987–89	Icewalk North Pole expedition; North Pole reached May 14, 1989; RS becomes first person to walk to both poles
1992	Keynote speaker at the first Earth Summit, Rio de Janeiro
1992–2002	Global mission: removal of 1,500 tons of trash from Antarctica; Local mission: partnered with African organization loveLife for AIDS-prevention presentations to more than 150,000 young people in South Africa
1996–97	One Step Beyond Expedition, South Pole to coast
2002	The overland journey of the sailboat *2041,* culminating at the World Summit in Johannesburg, South Africa
2003	Cape-to-Rio yacht race
2003–4	Circumnavigation of Africa in the *2041*
2004–5	Sydney Hobart Yacht Race with sails made of recycled plastic bottles
2003–present	Inspire Antarctic Expeditions: involving industry and young people in the preservation of Antarctica and the need for renewable energy
2008	E-Base—a pioneering education base in Antarctica powered by renewable energy and connected to schools worldwide via the Internet—goes live
	The Voyage for Cleaner Energy: campaigning for renewable energy in the sailboat *2041*

2009 (March)	E-Base goes live 365 days a year with an Internet connection powered by renewable energy
2009 (December)	Copenhagen Climate Change Conference with the sailboat *2041*
2012	The third World Summit for Sustainable Development
2041–48	The Madrid Protocol and moratorium on mining and drilling in Antarctica will be reviewed

Appendix 2

In the Footsteps of Scott Expedition, 1985–86

Date	Day	Distance (cumulative) in miles	Notes
Nov. 3	1	6.85 (6.65)	
Nov. 4	2	8.36 (15)	
Nov. 5	3	6.05 (21)	Bearing: 356°
Nov. 6	4	7.17 (28)	
Nov. 7	5	8.39 (36)	Bearing: 30°
Nov. 8	6	10.05 (46)	Bearing: 29°
Nov. 9	7	11.45 (52)	
Nov. 10	8	3.01 (61)	Blizzard
Nov. 11	9	13.25 (74)	Bearing: 28°
Nov. 12	10	11.84 (86)	Bearing: 27°
Nov. 13	11	6.77 (93)	Blizzard
Nov. 14	12	12.75 (105)	Wind force 2–3 SW–W; poor visibility in first five hours
Nov. 15	13	13.29 (119)	Good visibility: good surfaces except last 1 hr. 30 mins.
Nov. 16	14	11.58 (130)	Fresh snow cover blown off after first hour; breakable crust
Nov. 17	15	11.55 (142)	Surface hoar; breakable crust
Nov. 18	16	10.29 (152)	Poor contrast; soft crust
Nov. 19	17	11.64 (164)	Good contrast; soft slab; some settling
Nov. 20	18	12.22 (176)	Wind SE; good contrast; hard pulling on soft slab
Nov. 21	19	11.47 (187)	Poor contrast; soft slab

Date	Day	Distance (cumulative) in miles	Notes
Nov. 22	20	10.19 (198)	Zero contrast all day; some improvement of surface
Nov. 23	21	10.23 (208)	S wind force 3; good weather and contrast; bearing 25°
Nov. 24	22	6.75 (215)	Blizzard continues; fresh snow
Nov. 25	23	0.00 (215)	Did not move; no rations consumed
Nov. 26	24	13.04 (228)	Calm sunny day; good contrast; reasonable surface
Nov. 27	25	11.54 (239)	Good weather, visibility & surface; day cut short due to GW's feet. Bearing 24°
Nov. 28	26	12.55 (252)	Reasonable contrast; excellent surface; late start 11 A.M. Bearing 23°
Nov. 29	27	12.66 (264)	Good weather; no wind; sun; good contrast; reasonable surface
Nov. 30	28	12.14 (276)	1st session good contrast, warm; 2nd contrast less good; 3rd session contrast zero; snowing
Dec. 1	29	10.78 (287)	Reasonable surface & contrast; time out due to GW's feet
Dec. 2	30	10.96 (298)	Zero contrast all day; direction difficult; otherwise trouble-free
Dec. 3	31	12.73 (311)	Bearing 22°

Date	Day	Distance (cumulative) in miles	Notes
Dec. 4	32	8.96 (320)	GW calls halt; tent up; RS talks 1 hr. 30 min.; 6 P.M. blizzard; force 7–8
Dec. 5	33	0.00 (320)	No movement, wind S force 5; drifting snow
Dec. 6	34	16.65 (337)	Overcast all day; warm; wind calm; surface best yet; Mt. Hope visible
Dec. 7	35	16.07 (353)	Excellent surface again; GW tired; changed runner
Dec. 8	36	17.94 (371)	Calm sunny +4°C; good surface; GW's good day; downhill for last 3 miles
Dec. 9	37	16.75 (387)	Desolation Camp; easy day for all
Dec. 10	38	9.61 (397)	The Gateway: RS & GW wait at base; RM ascends
Dec. 11	39	0.00 (397)	Rest & recuperation at Gateway
Dec. 12	40	0.00 (397)	No movement: no rations consumed; climbed spur of Mt. Hope
Dec. 13	41	11.70 (409)	Calm, good visibility, little crevassing; camp opp. Monument Rock
Dec. 14	42	15.93 (425)	Easygoing day
Dec. 15	43	19.84 (444)	Blue ice; névé; sastrugi
Dec. 16	44	18.25 (463)	Sledge meters removed to prevent damage
Dec. 17	45	17.75 (480)	Last hour excellent
Dec. 18	46	20.77 (501)	Best day yet; new skins
Dec. 19	47	15.75 (517)	Scott's total Beardmore mileage 142 miles; ours 119.99 miles

Date	Day	Distance (cumulative) in miles	Notes
Dec. 20	48	15.38 (532)	Two sections of crevassing, no rope used. Bearing 57°
Dec. 21	49	16.74 (549)	Turned south, 6:15 P.M. Pressure ridges over?
Dec. 22	50	15.32 (564)	Surface, flat hard, wind-packed
Dec. 23	51	15.41 (580)	Crevassing 2nd session on ridge. Bearing: 10° mag.
Dec. 24	52	15.31 (595)	More ascent & long low waves
Dec. 25	53	15.57 (611)	Calm; good surface; long ascents
Dec. 26	54	17.00 (628)	Calm, warm; good surface, two shallow climbs
Dec. 27	55	17.28 (645)	Calm, warm; good surface; plateau ad infinitum
Dec. 28	56	17.22 (622)	Wind force 3–4; poor contrast; best session; more level
Dec. 29	57	16.51 (679)	Poor contrast; good surface; flat going. Bearing 9°
Dec. 30	58	16.96 (696)	Descent all day; last 3 hours heavy pulling; snow like dry sand
Dec. 31	59	16.02 (712)	No wind, beautiful day; new snow, 2' high sastrugi; 2nd & 3rd sessions very hard
Jan. 1	60	15.57 (727)	Sastrugi some 3.5' high. Continuing all 15 miles & very extensive & required finding a route through
Jan. 2	61	15.29 (742)	Sastrugi all day, diminishing in size but not extent; RM broken trace GW skins unstuck. Surface hoar

Date	Day	Distance (cumulative) in miles	Notes
Jan. 3	62	15.95 (758)	Hard day; diminishing sastrugi; better surface; much surface hoar
Jan. 4	63	8.33 (767)	Wind force 4; ground drift; rapid deterioration of visibility—100 m. Much ground drift; blizzard
Jan. 5	64	16.33 (783)	Wind force 3–4 gust 5; contrast poor-airborne snow; sastrugi ends; soft drift— very hard pulling, 20° parhelion
Jan. 6	65	16.22 (779)	Beautiful day, no wind, no cloud; heavy pulling in soft wind pack
Jan. 7	66	16.04 (815)	Bearing: 10°
Jan. 8	67	16.17 (831)	1st & 2nd sessions perfect, but heavy pulling in soft wind slab; 3rd— SW wind force 3
Jan. 9	68	10.79 (842)	1st session contrast intermittent; 2nd zero contrast & light snow; 3rd, abandoned. Bearing 5°
Jan. 10	69	16.21 (858)	W wind force 4; good contrast & visibility. Ground drift; deep soft pack, hard pulling 3rd session, waves & 2 long rises, saw aircraft in & out of Pole
Jan. 11	70	14.03 (872.99)	1st session abandoned— blizzard; 2nd sight flags; 3rd session arrived at Dome/Pole 11:53 P.M.

Appendix 3

Date	Day	Distance in miles
Mar. 20	1	5.3
Mar. 21	2	9.6
Mar. 22	3	4.6
Mar. 23	4	4.5
Mar. 24	5	6.3
Mar. 25	6	5.4
Mar. 26	7	No traveling
Mar. 27	8	No traveling
Mar. 28	9	No traveling
Mar. 29	10	No position for start of travel
Mar. 30	11	5.7
Mar. 31	12	7.7
Apr. 1	13	7.2
Apr. 2	14	2.5
Apr. 3	15	8.6
Apr. 4	16	10.8
Apr. 5	17	9.9
Apr. 6	18	5.1
Apr. 7	19	No traveling bad weather
Apr. 8	20	No traveling
Apr. 9	21	No traveling
Apr. 10	22	6.5
Apr. 11	23	12.8
Apr. 12	24	10.6
Apr. 13	25	26.2
Apr. 14	26	5.7
Apr. 15	27	8.1

Date	Day	Distance in miles
Apr. 16	28	9.0
Apr. 17	29	2.5
Apr. 18	30	5.8
Apr. 19	31	7.3
Apr. 20	32	11.0
Apr. 21	33	15.2
Apr. 22	34	13.8
Apr. 23	35	14.7
Apr. 24	36	13.8
Apr. 25	37	11.1
Apr. 26	38	7.2
Apr. 27	39	No travel
Apr. 28	40	No travel
Apr. 29	41	12.0
Apr. 30	42	15.5
May 1	43	14.7
May 2	44	12.8
May 3	45	15.6
May 4	46	11.5
May 5	47	16.0
May 6, 7	48	11.3
May 8	49	No travel
May 9	50	6.9
May 10	51	13.8
May 11	52	17.7
May 12	53	19.8
May 13, 14	54	10.9
May 15	55	Last position

Appendix 4

One Step Beyond Expedition, 1996–97

Date	Day	Distance in kilometers	Notes
Dec. 11	0	17	Parasail
Dec. 12	1	5	All man-haul
Dec. 13	2	51	Parasail
Dec. 14	3	13	Parasail
Dec. 15	4	30	Parasail
Dec. 16	5	—	No wind
Dec. 17	6	—	No wind
Dec. 18	7	33	Parasail
Dec. 19	8	60	Parasail
Dec. 20	9	55	Parasail
Dec. 21	10	35	Parasail
Dec. 22	11	53	Parasail
Dec. 23	12	29	Parasail
Dec. 24	13	23	Parasail
Dec. 25	14	—	No wind
Dec. 26	15	101	Parasail
Dec. 27	16	57	Parasail
Dec. 28	17	7	Parasail
Dec. 29	18	1	No wind
Dec. 30	19	—	No wind
Dec. 31	20	—	Waiting for aircraft
Jan. 1	21	46	Aircraft departs
Jan. 2	22	80	Parasail
Jan. 3	23	107	Parasail
Jan. 4	24	164	Parasail
Jan. 5	25	125	Parasail
Jan. 6	26	—	No wind

Date	Day	Distance in kilometers	Notes
Jan. 7	27	100	Parasail
Jan. 8	28	25	Two hours of wind
Jan. 9	29	16	One hour of wind
Jan. 10	30	2	Shear zone
Jan. 11	31	80	Parasail
Jan. 12	32	—	Headwind, RS departs
Jan. 13	33	—	Light headwind
Jan. 14	34	12	Parasail
Jan. 15	35	110	Parasail
Jan. 16	36	—	No wind
Jan. 17	37	102	Parasail
Jan. 18	38	118	Slowed by crevassing
Total	38 days	1,657 (1,030 m.)	

Parasailing averages

Days traveled	38
Days sailed	28
Average daily distance	44 km (27 m.) per day
Average daily sailed	60 km (37 m.) per day

Appendix 5

SS 2041 Ports of Call

Purchased May 14, 1999

Refit in the Netherlands, delivery to the United Kingdom
915 nautical miles

December 1999–November 2000

The Journey South
Plymouth, United Kingdom • Las Palmas, Spain • Recife, Brazil • Mar del
Plata, Argentina • Ushuaia, Argentina
9,212 nautical miles

November 2000

Delivery to Cape Town, South Africa—refit—delivery to Ushuaia, Argentina,
via Tristan da Cunha and the Falklands
8,846 nautical miles

December 2000–March 2001

Bellingshausen, Antarctica, to Cape Town, South Africa
4,423 nautical miles

April 2001–August 2002
The Overland Voyage, South Africa

Cape Town • Piketberg • Vanrhynsdorp • Upington • Prieska • Kimberley •
Vryburg • Mafikeng • Krugersdorp • Cullinan • Warmbaths • Pietersburg •
Louis Trichardt • Messina • Thohoyandou • Phalaborwa • Nelspruit •
Pongola • Richard's Bay • Durban • Port Shepstone • Kokstad • Umtata •
Quna • East London • Port Alfred • Port Elizabeth • Somerset East • Graaff

Reinet • Middelburg • Gariep Dam • Bloemfontein • Welkom • Sasolburg • Johannesburg • World Summit (September 2002–December 2002) • Return to Cape Town

7,000 nautical miles (12,000 km)

January 2003–March 2003
Cape to Rio Yacht Race—back to Cape Town for refit

7,420 nautical miles

April 2003–April 2004
Circumnavigation of Africa

Cape Town, South Africa, to Mauritius • Mombasa, Kenya • Dar es Salaam, Tanzania • Nacala • Beira, Mozambique • Maputo, Mozambique • Richard's Bay, South Africa • Durban, South Africa • East London, South Africa • Port Elizabeth, South Africa • Cape Town, South Africa • Walvis Bay, Namibia • Luanda, Angola • Accra, Ghana • Azores • Estepona, Spain • Sotogrande, Spain • Gibraltar, United Kingdom • Casablanca, Morocco • Tunisia • Crete, Greece • Port Said, Egypt • Suez, Egypt • Hurghada, Egypt • Eritrea • Djibouti • Cape Town, South Africa

23,500 nautical miles

April 2004–December 2004
Rolex Sydney Hobart Yacht Race: Cape Town, South Africa, to Sydney, Australia, via Melbourne

7,041 nautical miles
Sydney to Hobart, Tasmania
628 nautical miles

March 2008
Sydney, Australia, to Berkeley, California, via Ensenada, Mexico, and San Francisco, California

7,040 nautical miles

April 2008
Voyage for Cleaner Energy Phase 1, West Coast, United States
San Francisco, California • Seattle, Washington • Portland, Oregon •
San Diego, California

2,987 nautical miles

June 2008–July 2008

Delivery through the Panama Canal to Martha's Vineyard, Massachusetts, via Hurricane Alley (at the same time completing an eight-year global circumnavigation at latitude 38°6' S, longitude 68° 17' 31" W on July 24, 2008)
5,956 nautical miles

August 2008
Voyage for Cleaner Energy, Phase 2, East Coast, United States
Martha's Vineyard, Massachusetts • Nantucket, Massachusetts • New
York, New York • Washington, D.C., • Annapolis, Maryland •
Jacksonville, Florida

1535 nautical miles

Total: 86,511 nautical miles

Internet Resources

2041, the organization, with links to the *2041* (the boat), E-Base, itineraries, countdown clock, and other items of interest, including this list of resources updated (http://www.2041.com); 2041's Antarctica Curriculum (http://education.2041.com/)

CLIMATE CHANGE

Global Climate Change: NASA's Eyes on the Earth
   (http://climate.jpl.nasa.gov/Eyes.html)
United Nations Framework Convention on Climate Change
   (http://unfccc.int/2860.php)
Intergovernmental Panel on Climate Change (http://www.ipcc.ch/)
Act on CO_2, U.K. government website on climate change
   (http://www.direct.gov.uk/actonco2)
ACORE: American Council on Renewable Energy
   (http://www.acore.org/front)
An Inconvenient Truth (http://www.climatecrisis.net)
11th Hour Action (http://11thhouraction.com)

GOVERNMENT POLAR DIVISIONS

British Antarctic Survey (http://www.antarctica.ac.uk/)
U.S. Antarctic Program (http://www.usap.gov/)

GENERAL INTEREST

International Polar Year (http://www.ipy.org/)
Antarctica's Climate Secrets (http://www.andrill.org/flexhibit/)

Discovering Antarctica (http://www.discoveringantarctica.org.uk/)

World Wildlife Fund (http://www.worldwildlife.org/)

Cousteau Society (http://www.cousteau.org/)

Environmental Defense Fund (http://www.edf.org.cfm)

Sierra Club Clean Energy Solutions (http://www.sierraclub.org/energy/)

National Snow and Ice Data Center (http://nsidc.org/)

U.S. National Weather Service Marine Modeling and Analysis Branch, Sea Ice Analysis Page (http://polar.ncep.noaa.gov/seaice/Analyses.html)

The U.S. Geological Survey Satellite Map of Antarctica (http://terraweb.wr.usgs.gov/projects/Antarctica/AVHRR.html)

Cool Antarctica, a general website, with photos, maps, weather, and resources (http://www.coolantarctica.com/)